# Structure-and Adatom-Enriched Essential Properties of Graphene Nanoribbons

# Structure-and Adatom-Enriched Essential Properties of Graphene Nanoribbons

Shih-Yang Lin
Department of Physics, University of Houston

Ngoc Thanh Thuy Tran
Department of Physics, National Cheng Kung University, Taiwan

Sheng-Lin Chang
Department of Physics, National Chiao Tung University, Taiwan

Wu-Pei Su
Department of Physics, University of Houston

Ming-Fa Lin
Quantum Topology Center, National Cheng Kung University,
Taiwan

CRC Press
Taylor & Francis Group
Boca Raton London New York

CRC Press is an imprint of the
Taylor & Francis Group, an **informa** business

CRC Press
Taylor & Francis Group
6000 Broken Sound Parkway NW, Suite 300
Boca Raton, FL 33487-2742

First issued in paperback 2020

© 2019 by Taylor & Francis Group, LLC
CRC Press is an imprint of Taylor & Francis Group, an Informa business

No claim to original U.S. Government works

ISBN 13: 978-0-367-65690-4 (pbk)
ISBN 13: 978-0-367-00229-9 (hbk)

**Visit the Taylor & Francis Web site at**
**http://www.taylorandfrancis.com**

**and the CRC Press Web site at**
**http://www.crcpress.com**

# Contents

Authors                                                                  vii

Preface                                                                   ix

Contributors                                                              xi

1  Introduction                                                            1

2  The first-principles method and experimental instruments              11
   2.1  Theoretical calculations  . . . . . . . . . . . . . . . . . . . .  11
   2.2  Experimental instruments  . . . . . . . . . . . . . . . . . . . .  14

3  Monolayer graphene nanoribbons                                        25
   3.1  Armchair systems  . . . . . . . . . . . . . . . . . . . . . . . .  27
   3.2  Zigzag systems  . . . . . . . . . . . . . . . . . . . . . . . . .  35

4  Curved and zipped graphene nanoribbons                                49
   4.1  Curved nanoribbons  . . . . . . . . . . . . . . . . . . . . . . .  50
   4.2  Carbon nanotubes  . . . . . . . . . . . . . . . . . . . . . . . .  59

5  Folded graphene nanoribbons                                           71
   5.1  The rich geometric structures  . . . . . . . . . . . . . . . . .  72
   5.2  The unusual electronic and magnetic properties  . . . . . . . .  77

6  Carbon nanoscrolls                                                    87
   6.1  The optimal geometries  . . . . . . . . . . . . . . . . . . . . .  88
   6.2  Electronic properties and magnetic configurations  . . . . . .  93
   6.3  Comparisons among the planar, curved/zipped, folded, and
        scrolled systems  . . . . . . . . . . . . . . . . . . . . . . . .  99

7  Bilayer graphene nanoribbons                                         103
   7.1  Stacking-enriched  geometric  structures  and  magnetic
        configurations  . . . . . . . . . . . . . . . . . . . . . . . . . 104
   7.2  Diverse electronic properties  . . . . . . . . . . . . . . . . . 110
   7.3  Differences between bilayer 1D and 2D systems  . . . . . . . . 114

**8   Edge-decorated graphene nanoribbons**                             **123**
   8.1   Adatom edge passivations . . . . . . . . . . . . . . . . . . . . .   124
   8.2   Decoration- and curvature-enriched essential properties  . . .   134

**9   Alkali-adsorbed graphene nanoribbons**                            **147**
   9.1   The alkali-created conduction electrons  . . . . . . . . . . . .   148
   9.2   The    edge-    and    adsorption-co-dominated    magnetic
          configurations   . . . . . . . . . . . . . . . . . . . . . . . . . .   157

**10  Halogen-adsorbed GNRs**                                          **163**
   10.1  Fluorination effects   . . . . . . . . . . . . . . . . . . . . . . .   164
   10.2  Chlorination-related systems   . . . . . . . . . . . . . . . . . .   176

**11  Metal-adsorbed graphene nanoribbons**                           **189**
   11.1  Al  . . . . . . . . . . . . . . . . . . . . . . . . . . . . . . . . . .   192
   11.2  Ti  . . . . . . . . . . . . . . . . . . . . . . . . . . . . . . . . . .   201
   11.3  Bi  . . . . . . . . . . . . . . . . . . . . . . . . . . . . . . . . . .   212
   11.4  Fe/Co/Ni . . . . . . . . . . . . . . . . . . . . . . . . . . . . . .   225

**12  Concluding remarks**                                             **233**

**References**                                                        **245**

**Index**                                                            **279**

# *Authors*

**Shih-Yang Lin** received his PhD in physics in 2015 from the National Cheng Kung University (NCKU), Taiwan. Since 2015, he has been a postdoctoral researcher at NCKU. Recently, he is a visitor scholar in University of Houston. His research interests include low-dimensional group IV materials and first-principle calculations.

**Ngoc Thanh Thuy Tran** obtained her PhD in physics in January 2017 from the National Cheng Kung University (NCKU), Taiwan. Since 2017, she has been a postdoctoral researcher at NCKU. She is currently working at Hierarchical Green-Energy Materials (Hi-GEM) Research Center, NCKU. Her scientific interests are focused on the functionalization of graphene and its derivatives using first-principle calculations.

**Shen-Lin Chang** obtained his PhD in 2014 in physics from the National Cheng Kung University (NCKU), Taiwan. Since 2014, he has been a postdoctoral researcher at NCKU and NCTU. His main scientific interests are in the field of condensed matter physics. Most of his research is focused on the electronic and magnetic properties of one-dimensional nanomaterials.

**Wu-Pei Su** is a professor in the Department of Physics, University of Houston. His research interest has been focusing on solving the X-ray phase problem in protein crystallography. It is a problem encountered in constructing three-dimensional molecular images of proteins from the X-ray diffraction data of protein crystals. For protein crystals with high solvent content, Su's team has discovered a solution to the problem. He is currently extending the solution to a more general crystal.

**Ming-Fa Lin** is a distinguished professor in the Department of Physics, National Cheng Kung University, Taiwan. He received his PhD in physics in 1993 from the National Tsing-Hua University, Taiwan. His main scientific interests focus on essential properties of carbon related materials and low-dimensional systems. He is a member of American Physical Society, American Chemical Society, and the Physical Society of Republic of China (Taiwan).

# Preface

This book aims to provide a systematic review of the feature-rich essential properties in emergent graphene nanoribbons, covering both up-to-date main-stream theoretical and experimental researche. It includes a wide range of 1D systems, namely, armchair and zigzag graphene nanoribbons without/with hydrogen terminations, curved and zipped graphene nanoribbons, folded graphene nanoribbons, carbon nanoscrolls, bilayer graphene nanoribbons, edge-decorated graphene nanoribbons, and alkali-, halogen-, Al-, Ti, and Bi-absorbed graphene nanoribbons. The first-principles calculations are successfully developed to thoroughly explore the physical, chemical, and material phenomena; furthermore, the concise pictures are proposed to explain the fundamental properties. The systematic studies on the various effects, which are due to edge structure, quantum confinement, curing, zipping, scrolling, folding, stacking, layer, orbital, spin, and adatom chemisorption, can greatly promote the basic and applied sciences. They could attract much attention from researchers in the scientific community, not only for the study of emergent 1D materials, but also for the exploration of other 0D, 2D, and 3D systems.

The content presents the reliable and complete calculation results, the other theoretical models, and the various experimental measurements and applications. The optimal geometric structures, electronic properties, and magnetic configurations are investigated in detail by the first-principle density functional theory using the Vienna ab initio simulation package (VASP). The exchange-correlation energy due to the electron-electron interactions is evaluated from the Perdew-Burke-Ernzerhof functional under the generalized gradient approximation. The projector-augmented wave pseudopotentials are employed to evaluate the electron-ion interactions. The spin configurations are taken into account for any graphene nanoribbons, especially for the adatom-adsorbed systems. Moreover, the van der Waals force is employed to correctly describe the interactions between two graphene layers. The calculated results cover the ground state energy, binding energy, non-uniform honeycomb lattice, planar/non-planar structure, interlayer distance, bond length, bond angle, position and height of adatom, adatom nanostructure, band structure, width-dependent band gap, adatom-induced energy gap and free carriers, charge distribution, spin configuration, magnetic moment, and orbital- and spin-decomposed DOSs.

Specifically, the critical orbital hybridizations are accurately examined from the atom-dominated energy bands, the spatial charge density and the difference after chemisorption/bonding, and atom- and orbital-projected DOSs.

Furthermore, the diverse magnetic configurations are delicately identified from the strong competition between zigzag edge carbons and adatoms, the spin-split/spin-degenerate band structure, the spin-induced net magnetic moment, and the spatial spin distributions around host atoms and guest adatoms. Such theoretical framework is absent in the published books and papers of other research groups, and it could be generalized to emergent 0D-3D materials. A detailed comparison with the other theoretical studies and the experimental measurements is made. The high potentials in various materials applications are also discussed.

# *Contributors*

**Shih-Yang Lin**
Department of Physics
University of Houston, USA
Department of Physics
National Chung Cheng University,
Taiwan

**Ngoc Thanh Thuy Tran**
Hi-GEM
National Cheng Kung University, Taiwan
Department of Physics
National Cheng Kung University, Taiwan

**Sheng-Lin Chang**
Department of Physics
National Chiao Tung University, Taiwan

**Wu-Pei Su**
Department of Physics
University of Houston, USA

**Ming-Fa Lin**
Quantum Topology Center
National Cheng Kung University, Taiwan
Hi-GEM
National Cheng Kung University, Taiwan

# 1

## *Introduction*

Carbon atoms, which possess four active valence electrons, can create the dimension- and curve-related condensed-matter systems, such as, graphite [26, 342], graphene [255, 103], graphene nanoribbon (GNR) [193, 117], carbon nanotube [139, 140], carbon toroid [220, 300], and fullerene [170, 169]. The carbon-related materials, with the diverse chemical bondings ($sp^3$, $sp^2$, and $sp$ bondings), have attracted a lot of theoretical and experimental researche since the successful syntheses of $C_{60}$ in 1985 [170], multi-walled carbon nanotubes in 1991 [139], carbon tori in 1998 [220], and few-layer graphenes in 2004 [255]. They are very suitable for studying the novel physical, chemical, and material phenomena, especially for the emergent low-dimensional systems. GNRs are one of the main stream nanomaterials, because the essential properties are greatly diversified by the complex relations among honeycomb lattice, one-atom thickness, finite-size quantum confinement, edge structure, planar/non-planar geometry, stacking configuration, adatom decoration, and chemical doping. Each GNR could be regarded as a finite-width graphene or an unzipped carbon nanotube. Up to now, GNRs have been successfully synthesized by the various experimental methods under the top-down and bottom-up schemes. The direct cutting of graphene layers covers the lithographic patterning and etching [117, 17], sonochemical breaking [193, 355], metal-catalyzed cutting [81, 76], and oxidization reaction [238, 100]. The unzipping of multi-walled carbon nanotubes is achieved by the chemical attack [166, 51], laser irradiation[171], plasma etching [147, 146], metal-catalyzed treatment [93, 264], hydrogen treatment and annealing [332], scanning tunneling microscope (STM) tip [261], transmission electron microscopy (TEM) [155], intercalation and exfoliation [47, 167], and electrochemical [310] and sonochemical processes [145, 369]. Moreover, other strategies involve chemical vapor deposition [45, 361, 325] and chemical synthesis [377, 44, 33], being able to produce macroscopic quantities of graphene nanoribbons. This systematic work is focused on the width-, layer number-, stacking-, curving-, folding-, scrolling-, decorating-, and doping-dependent geometric structures, electronic properties, and magnetic configurations which are mainly investigated by using the first-principles calculations. Monolayer and bilayer GNRs, curved GNRs, carbon nanotubes, folded GNRs, carbon nanoscrolls, adatom-passivated GNRs, and alkali-, halogen-, and (Al,Bi,Ti)-doped GNRs are taken into account. We propose physical and chemical pictures to fully comprehend the rich and unique phenomena. The single- and multi-orbital hybridizations between the same or distinct atoms, and the edge- and adatom-enriched spin configurations are thoroughly analyzed from the calculated results. Detailed comparisons with the other theoretical predictions and the experimental measurements are made. The high potentials of various materials applications are also discussed.

Graphene nanoribbons, with one-dimensional quantum confinement effects, can greatly diversify the essential properties, shedding light on potential applications in nanodevices, and thus attracting numerous theoretical studies. Monolayer graphene is a zero-gap semiconductor with zero density of states (DOS) at the Fermi level, and the other layered graphenes are semi-metals because of the interlayer atomic interactions. It should be noticed that the vanishing energy gaps have induced high barriers in certain important applications. However, electronic and magnetic properties of graphene nanoribbons are strongly affected by the ribbon widths [323], crystallographic orientations [21], and edge saturations [180]. Energy gaps, which are created by the finite-size effects, are confirmed in both armchair and zigzag GNRs. That is, they are found to be very sensitive to the change in edge structure. In addition to the parabolic energy bands, a pair of partially flat valence and conduction bands crossing the Fermi level is predicted to only exist in zigzag nanoribbons. The tunable electronic properties have further stimulated a lot of researche on magnetic properties [324, 91], optical excitations [376, 136], and transport properties [197, 350]. Experimentally, the width dependent energy gaps are verified by the measurements of electrical conductance [117]. The atomic structures, 1D parabolic energy dispersions, and low-lying asymmetric/symmetric peaks of the density of states are verified by STM [336], angle-resolved photoemission spectroscopy (ARPES) [294], and scanning tunneling spectroscopy (STS) [160], respectively. On the application side, GNRs have been explored and utilized in semiconductor industry, environmental engineering, and biochemistry. GNRs with width below 10 nm can serve as field-effect transistors (FETs) at room temperature [193, 355]. GNR arrays down to 5 nm width exhibit an exceptional $NO_2$ sensing performance [1]. A GNR transistor integrated with solid-state nanopore becomes a sensor for DNA translocation [341]. How many kinds of energy gaps with/without edge passivations and their origination are the core topics of this systematic study.

The changes in geometric structures of pristine graphene nanoribbons can modify and enrich the essential properties. The curved systems are directly synthesized from the cylindrical carbon nanotubes using physical and chemical methods [166, 51, 171, 147, 146, 93, 264, 332, 261, 155, 47, 167, 310, 145, 369]. The unzipping of multi-walled carbon nanotubes offers an efficient way in producing graphene nanoribbons with controlled structure and quality. The unzipped and zipped graphene structures have been observed under the measurements of atomic force microscopy (AFM), TEM, and scanning electron microscopy (SEM). The curved surfaces and dangling bonds are an ideal chemical environment for modulating the geometric structure and electronic properties. A planar graphene nanoribbon presents a typical $sp^2$ bonding, being determined by the $\pi$ bonding of the parallel $2p_z$ orbitals and the $\sigma$ bonding of the in-plane $(2s, 2p_x, 2p_y)$ orbitals. On the other hand, a perfect cylindrical surface in a carbon nanotube can create the coexistent $sp^2$ and $sp^3$ hybridizations, as verified by theoretical calculations [150, 314] and experimental measurements [260]. Apparently, graphene nanoribbons and carbon nanotubes sharply contrast with each other in the curvature and boundary condition, and so do the other essential properties. A cylindrical system has a periodical boundary condition along the azimuthal direction, so that the angular momentum serves as a good quantum number in addition to the axial wave vector. The transferred longitudinal momentum and transverse angular momentum are conserved in various scattering mechanisms, e.g., optical excitations [4, 210, 313] and Coulomb interactions [226, 209, 208]. However, an open boundary condition in a finite-width nanoribbon cannot create the transverse quantum number, leading to the absence of its conservation law in the excitation processes [75, 303, 304, 41]. These two critical factors, being responsible for the essential properties of the curved systems, are explored in details, such as the effects on the ground state energy, the effective interaction distance, the bond length, orbital hybridization, the dominance of edge carbon atoms, and the semi-conducting/metallic behavior.

The folding of a honeycomb lattice may induce new and distinct properties in graphene-related systems. The folded graphene nanoribbons have been produced in experimental laboratories [193, 224, 387, 219, 135], and have been confirmed by the direct mapping of the open and closed edges using the high-resolution TEM (HRTEM). Transport measurements show that they could act as high-performance FETs [193] and electrode materials [60], Each unique geometric structure is composed of a chiral/non-chiral open boundary, a bilayer-like graphene, and a half carbon nanotube; that is, it consists of three distinct carbon honeycomb regions. For example, two zigzag/armchair graphene nanoribbons are connected by a fraction of armchair/zigzag carbon nanotube, in which the former are stacked in the manner of AA, AB or other configurations. The folded systems can provide rich physical and chemical environments in terms of edge structure, finite size, curvature, and orbital bonding. They are predicted to exhibit unusual phenomena, such as the magnetic quantization of highly degenerate Landau levels [278, 97], the semiconductor-metal transition under a structural transformation [60], the complex width dependences of energy gaps, and the diverse spin configurations. The combined effects, which arise from the quantum confinement, the spin arrangement near the specific edges, the edge-edge interactions, the van der Waals interactions between two planes, and the mechanical bending of a curved surface, will play critical roles in the optimal geometries, electronic properties, and magnetic configurations. This is worthy of a systematic investigation by the first-principles calculations.

Carbon nanoscrolls, with the spiral cross-sections, were first reported by Bacon in 1960 using the arc-discharge of graphite [16]. The distinct synthesis methods are further developed to fabricate high-quality carbon nanoscrolls more efficiently [189, 345, 305, 311, 315, 370]. They present tube-like structures under the disclosed surfaces. Each nanoscroll is an open-ended spirally wrapped graphene nanoribbon. The main features cover the inner and outer edges, the non-uniform curved surfaces, and the multi-walled structures with tunable interlayer distances. The formation and structure stability could be simulated by the molecular dynamics method [40]. As a result of the unusual structures, the essential properties are easily modulated by chemical doings, external fields, and mechanical strains. Carbon nanoscrolls are expected to have high potentials in applications, such as hydrogen storages [246], electromechanical nanoactuators [295], FETs, and microcircuit interconnect components [370]. First-principles calculations on energy bands shows strong dependence on edge structure and ribbon width [69, 172], as revealed in planar systems. However, the optimal structures, the multi-orbital hybridizations, and the spin arrangements need to be taken into consideration to fully understand the unusual geometric, electronic, and magnetic properties, e.g., the critical configuration parameters, the diverse width dependences of energy gaps, and the specific spin configurations (Chapter 6). Apparently, the structure-induced combined effects are quite different from among the scrolled, folded, curved, and planar graphene nanoribbons. This means that the geometric structures will result in diverse essential properties.

Few-layer graphene nanoribbons are frequently produced in various experimental syntheses [17, 193, 166, 147, 361]. The stacking symmetries of layered systems play critical roles in diversifying the essential properties, as confirmed by many theoretical [202, 340, 357, 71, 165, 265] and experimental [258, 257, 316, 38, 327, 107, 77, 37, 329] studies on 2D few-layer graphenes. For example, the trilayer graphenes, with AAA, ABA, ABC, and AAB stackings, present the linear, parabolic, sombrero-shape, and oscillatory for the first pair of energy bands measured from the Fermi level. The dramatic transitions will come to exist by modulating the stacking configurations between two layers, such as the metal-semiconductor transition and the generation/destruction of magnetism [58]. The shift-dependent variation of stacking symmetries could be achieved by electrostatic manipulation of STM, as done for graphene-related systems [373, 98, 225, 282]. Specifically, bilayer zigzag graphene nanoribbon could serve as a model system in revealing the combined effects due to the edge-edge interactions, the stacking configurations, and the intralyer and interlayer spin arrangements. Previous theoretical calculations on typical AA and AB stackings predict the planar/non-planar profiles and the magnetic/non-magnetic configurations [111, 301, 201, 266]. More complete calculated results are required to clearly identify the stacking-induced diverse behaviors. The shift-dependent optimal geometries, relatively stable stacking configurations, magnetic moments, spin distributions, spin degeneracies, low-lying energy dispersions, band gaps/free carrier densities, spatial charge densities near two open edges, and van Hove singularities are worthy of further detailed examinations. Furthermore, theoretical studies on the sliding bilayer graphene can fully explain the confinement-induced effects and the dimensional crossover.

How to dramatically change the essential properties is one of the main-stream topics in materials science and application. The chemical function-alization is one of the most effective ways. Pristine graphene nanoribbons, without passivation, possess a honeycomb lattice and two open edges, provide a reactive chemical environment. The intrinsic dangling bonds associated with edge carbon atoms are highly unstable, so that the dopants could be applied to passivate the active boundary. The dopant-created chemical bonding will play a critical role in determining the most stable configuration and thus other essential properties. Up to now, the H- [330], Cl- [333], and O-terminated [166] graphene nanoribbons have been produced by chemical synthesis methods, as examined by the measurements of STM, X-ray diffraction, and XPS. The first and the second adatom-decorated systems are, respectively, identified to present the planar and distorted edge structures with an open boundary. According to the theoretical calculations, the active edges tend to adsorb atoms [234, 133, 228, 131, 268, 105, 302, 14, 356, 317, 348, 217], molecules [52, 307, 80] or radical groups [52, 307]. The predicted decoration effects cover the edge reconstructions, the adatom-dominated energy bands, the greatly reduced energy gaps, and the semiconductor-metal transitions. They are closely related to the strong competitions between the edge passivation and the finite-size confinement. How many types of edge structures and width-dependent energy gaps can exist needs to be thoroughly clarified from detailed calculations and discussions. Specially, the adatom/molecule decorations might happen during the unzipping of carbon nanotubes through strong chemical reactions. The edge passivation and curvature are expected to have strong effects on the geometric, electronic, and magnetic properties [57], such as the adatom- and curvature-dominated planar/curved/tubular structures, the metallic/semiconducting behaviors, and the anti-ferromagnetic/ferromagnetic/non-magnetic configurations.

Adatom adsorptions on honeycomb surfaces can greatly modify the essential properties of graphene nanoribbons. Alkali, halogen, and metal atoms (Chapters 9–11), which possess the unique chemical and physical features, are very suitable for exploring the adsorption-induced dramatic changes, e.g., the semiconductor-metal transition, the n- or p-type doping (the creation of free electrons or holes), and the transformation between different spin arrangements. Up to now, alkali atoms have been successfully synthesized in carbon-related systems, such as graphite [360], fullerene [267], carbon nanotube [39], and graphene [128]. The alkali-doped systems are verified to present metallic behavior as a result of the high-density conduction electrons. They have displayed potential applications, e.g., the Li-ion batteries based on graphene [191]. There are many theoretical studies on the n-type doping effects [104, 271]. The alkali-adsorbed graphene and graphene nanoribbon are predicted to exhibit the blue-shift Fermi level [218]. The former is confirmed by the ARPES measurements [398]. It is well known that each alkali atom has one very active s-orbital, leading to significant chemical bonding with carbon atom. The critical orbital hybridization, which is determined by the delicate energy bands, spatial charge distribution, and DOS (discussed later in Section 2.1), is expected to dominate the diversified electronic properties and magnetic configurations. Determination of the number of conduction electrons due to the outmost s-orbitals and fine tuning of the spin magnetism are the main focuses under systematic investigations.

Halogen atoms are very good candidates for chemical modifications, mainly owing to the rather strong electron affinities. They have five electrons in the outmost $(p_x, p_y, p_z)$ orbitals; furthermore, the adatom-induced spin states might come to exist after chemical adsorptions. The halogenated graphene nanoribbons are expected to exhibit complicated orbital hybridizations in chemical bonds and the adatom- and edge-C-created magnetic configurations. The diverse electronic and magnetic properties will be present in the variations of concentrations, distributions, and edge structures. How to identify the charge- and spin-dependent mechanisms is a critical issue and needs to be examined from the various physical quantities. Recently, there are many experimental [277, 292, 185] and theoretical [240] studies on the halogenation effects of graphene-related systems. Fluorinated [292], chlorinated [185], brominated [194] and iodinated [389] graphene systems have been successfully synthesized by various chemical methods. The important differences between fluorinated and other halogenated graphene systems in the semiconducting or metallic behavior are confirmed by the transport [365], optical [61], and ARPES measurements [352]. However, only very few studies are conducted on 1D halogenated graphene nanoribbons. Theoretical calculations predict very strong fluorination effects, such as the multi-orbital hybridizations in F-C, C-C, and F-F bonds, the irregular relations between energy gaps (valence hole densities) and adatom-adsorption cases, and five kinds of electronic and magnetic properties [253]. All the halogenation effects arising from the distinct kinds of adatoms are included in this work, especially the adsorption conditions/mechanisms in determining the feature-rich energy bands, free carrier densities, magnetic/NM configurations, and significant differences of the F-C and (Cl,Br,I,At)-C bonds.

Metal atoms, including aluminum, titanium, bismuth, and iron/cobalt/nickel could provide effective dopings in conduction electrons (the *n*-type dopings) [216, 66, 64, 63], as revealed in alkali ones. After adsorption on graphene nanoribbon surfaces, they are expected to induce more complicated multi-orbital hybridizations in the significant carbon-metal bonds, compared with the $2p_z$ orbital and the outermost $s$ orbital in the carbon-alkali bonds. Based on detailed first-principles calculations on the Al-, Ti/Fe/Co/Ni-, and Bi-adsorbed graphenes, the critical bondings are identified to arise from the $(3s, 3p_x, 3p_y)$, $(3d_{z^2}, 3d_{xy}, 3d_{x^2-y^2})$, and the $(6s, 6p_x, 6p_y, 6p_z)$ orbital, respectively. The theoretical predictions could highly promote further understanding of experimental measurements. For example, the aluminum-based batteries have been developed quickly to greatly enhance the charging and discharging reactions and to reduce the cost of the metallic anode [206]. The Al-adsorbed graphene is predicted to create as many free carriers as the alkali adatoms do, while the concentration of the former has an upper limit of 25% [216]. When the low-coverage titanium adatoms are adsorbed on monolayer graphene supported by 4H-SiC(0001) substrate, the high-density free electrons are verified from the high-resolution ARPES measurements [66]. The theoretical calculations suggest that the Ti adsorptions could exhibit the high-concentration adsorptions. The measured results agree with the calculated ones under the very strong orbital hybridizations in the Ti-C bonds. Specially, a large-scale hexagonal array of Bi adatoms on graphene surface is clearly revealed at room temperature using the STM measurements [64, 63]. It becomes three- and four-member nanostructures under the annealing process. The six-layer SiC substrate, the buffer graphene layer, and the slightly deformed monolayer graphene in the first-principles model are proposed to explain the most and meta-stable optimal structures [213]. The critical orbital hybridizations in generating free conduction electrons are also examined in detail [213]. Four different stable structures have been observed on Fe adsorbed graphene, and the corresponding DFT calculations shows that these structures all possess magnetic moments of either 2.00 or 4.00 $\mu$B [120]. The adsorption site, magnetic ground state, and anisotropy of single Co adatoms on 2D graphene are determined by experiments [88]. The different adatom concentration, the metal-induced free carrier density, and the magnetic properties are focuses of the metal-adsorbed graphene nanoribbons.

A thorough and systematic study is conducted on the essential properties of the structure-enriched and adatom-doped graphene nanoribbons. It covers a lot of critical mechanisms, lattice symmetries, boundary conditions, quantum confinements, edge structures, stacking-dependent interlayer interactions, curvature effects, edge-edge interactions, distributions and concentrations of adatoms, charge transfers, orbital hybridizations, magnetic environments, and adatom-created moments/spin configurations. The optimal geometric structures are related to the most/relatively stable configurations, various non-planar profiles, interlayer and edge-edge distances, stacking configurations, critical curvatures, minimum widths/inner diameters, saturated nanotube diameters, bond lengths, bond angles; positions and heights of adatoms. Energy bands are characterized by the dominances of carbon, adatom and (carbon, adatom), linear, parabolic, partially flat and oscillatory energy dispersions, symmetry or asymmetry about the Fermi level, energy gaps due to the combined effects, free electron/hole densities, state degeneracy, and spin-related splitting. The 1D critical points in energy bands reveal as the special van Hove singularities, including plateaus, asymmetric peaks, and symmetric peaks. Specifically, the delicate atom- and orbital-projected DOSs, the atom-dominated energy bands and the spatial charge densities, provide much information about the dominating chemical bondings. They could be used to identify the orbital hybridizations in C-C, C-X, and X-X bonds (X adatom). Furthermore, the magnetic configurations, the ferro-magnetic, anti-ferro-magnetic, and non-magnetic ones (FM, AFM, and NM) are examined from the net magnetic moment, the spin-dependent energy bands, the spatial spin distribution, and the spin-decomposed DOSs. According to the rich and unique essential properties, materials functionalities and potential applications are further discussed.

This book is organized as follows. Chapter 2 includes a detailed discussion of the computational methods and experimental tools, as well as important progresses achieved recently. The geometric, electronic, and magnetic properties are studied for monolayer systems in Chapter 3, especially for the finite-size and boundary effects. Chapter 4 focuses on the zipped/unzipped variations of curved systems, in which the critical arc angles, the edge-induced chemical bondings and energy bands, and the semiconductor-metal transitions are explored thoroughly. A detailed comparison is made between curved nanoribbons and carbon nanotubes. In Chapter 5, eight typical types of high-symmetry folded graphene nanoribbons are considered for studying the bilayer- and tuble-like composite structures, the unusual low-lying energy dispersions, the semiconductor-metal transitions, the diversified width dependences of band gaps, and the distinct magnetic environments, being based on the complex combined effects. How to sustain a stable carbon nanoscroll with an optimal spiral profile is discussed in Chapter 6, especially for the sufficiently large ribbon width and inner diameter. The edge-, width-, and spin-created diverse properties are investigated. Furthermore, the important differences among the planar, curved, folded, and scrolled graphene nanoribbons are fully clarified. Chapter 7 presents the dramatic transformations of the essential properties in bilayer zigzag graphene nanoribbon caused by the relative displacement between two layers. Also discussed is sliding bilayer graphene to illustrate the dimension-dominated behaviors. How to modify the boundary structure with the chemical edge passivation is the focus of Chapter 8. The decoration and curvature effects are examined for a lot of symmetric configurations and various adatoms to explore the diversified essential properties, especially for the types of geometric structures and the metallic/semiconducting behaviors.

The surface adatom adsorptions are investigated for graphene nanoribbons with hydrogen edge passivation. The critical chemical bondings and the adatom-induced spin states are proposed to explain the diversified properties. Chapter 9 explores the alkalization effects on the geometric, electronic, and magnetic properties. Whether there exists a specific relation between the alkali concentration and the 1D conduction electron density is examined for any absorption configurations. Furthermore, the spin arrangements at zigzag edges are delicately tuned by changing the position and concentration of alkali adatoms. The fluorination and chlorination-related effects are investigated in Chapter 10, especially for the significant mechanisms. The focus is on how to achieve the planar/buckled honeycomb structure, the high/low binding energy, the destruction/existence of the $\pi$-electronic band structure, the halogen-dependent energy bands, and the metallic or semi-conducting behavior through the multi-/single-orbital hybridizations in X-C bonds. Furthermore, the adsorption-dominated spin configuration and their strong competition with the zigzag-edge one are expected to create different types of magnetic configurations. Chapter 11 thoroughly discusses the main features of the Al-, Ti-, and Bi-adsorbed graphene nanoribbons and the significant differences among such systems. It includes the complicated orbital hybridizations arising from the distinct (s,p,d) orbitals, the metal-dominated low-lying conduction and valence bands, the adatom-induced free electron densities, and the adatom-dependent spin distributions. Finally, Chapter 12 includes concluding remarks and future studies.

# 2

# The first-principles method and experimental instruments

The details of numerical evaluations and experimental measurements are discussed in the following two sections.

## 2.1 Theoretical calculations

A typical condensed-matter crystal consists of a periodic atomic arrangement, in which each atom might have several valence electrons distributed around an ionic core. Any system presents the complicated many-body effects due to the electron-electron Coulomb interactions and the electron-ion crystal potential. Apparently, it is rather difficult to accurately deal with the many-particle Schrodinger equation. The evaluation difficulties will be greatly enhanced when the unusual geometric structures and the strong chemical adsorptions need to be taken into consideration, e.g., the folded/scrolled and adatom-doped graphene nanoribbons. Some approximation methods have been developed to obtain the reliable electronic states. Up to now, the first-principles calculations become a dominant method in getting the quantum states of periodic systems. Such numerical calculations are very efficient for studying the essential physical properties. Specifically, Vienna *ab initio* simulation package (VASP) [168] evaluates an approximate solution within the density functional theory by solving the Kohn-Sham equations. [163] Electron charge density is responsible for all the interactions of a periodic condensed-matter system; that is, it can determine the ground state energy and the essential properties. The spatial charge density will be solved by the numerical self-consistent scheme, as indicated in a flow chart of calculation (Figure 2.1). Compared to this numerical method, the tight-binding model, with the atomic interaction parameters, cannot not be utilized to investigate the optimal geometric structures and the complicated orbital hybridizations in various chemical bonds. However, this model is efficient in studying the essential properties under the external fields, e.g., the quantized Landau levels [109, 232, 173, 75, 203], magneto-optical excitations [112, 113, 126], magnetoplasmons [24, 384], Hall quantum transports [151, 368]. and transverse electric-field-induced semiconductor-meta transitions [239]. In addition, the direct experimental verifications on the Landau wave functions of layered graphene systems are absent up to date, while they could be identified from the STM/STS examinations for few-layer graphene nanoribbons ( the well-behaved quasi-Landau levels with the oscillatory and localized spatial probability distributions under the specific number of zero points).

The geometric structures, electronic properties, and magnetic configurations are thoroughly studied for 1D GNR-related systems using VASP [168]. The spin-polarized density functional theory is used to realize the magnetic configurations arising from the edge structures and adatom adsorptions [270]. The Perdew–Burke–Ernzerhof functional under the generalized gradient approximation can characterize the many-body Coulomb interactions, the exchange and correlation energies [270]. The projector-augmented wave pseudopotentials account for the electron-ion interactions [35]. In solving the many-particle Schrodinger equation, plane waves, with a maximum energy cutoff of 400 eV, are the bases in building the Bloch wave functions. The 1D periodic boundary condition is along $\hat{x}$, and the vacuum spacing associated with $\hat{y}$ and $\hat{z}$ is larger than 15 Å to avoid the interactions between two neighboring nanoribbons. The Brillouin zone is sampled by $15 \times 1 \times 1$ and $600 \times 1 \times 1$ k point meshes within the Monkhorst–Pack scheme for geometric optimizations and further calculations on electronic structures, respectively. The convergence of energy is set to be $10^{-5}$ eV between two simulation steps, and the maximum Hellmann–Feynman force acting on each atom is less than 0.01 eV/Å during the ionic relaxations. Specifically, to correctly describe the significant atomic interactions between two neighboring graphene nanoribbons, the van der Waals force is included in the calculations by the semi-empirical DFT-D2 correction of Grimme [106].

For the structure-enriched and adatom-doped GNRs, the theoretical calculations cover the ground state energy, binding energy, interlayer distance, bond length, bond angle, planar/curved/buckled/scrolled/folded/stacked honeycomb lattice, hexagonal/non-hexagonal edge decoration, position and height of adatom, atom-dependent band structure, adatom-induced carrier density and energy gap, spatial charge distribution, spin arrangement, magnetic moment, and orbital- and spin-projected DOSs. The detailed analyses on the calculated results can get the concise physical and chemical pictures and thus fully comprehend the cooperative/competitive relations among the honeycomb lattice, the finite-size quantum confinement, the edge structure, the interlayer atomic interaction, the non-planar curvature effect, and the critical orbital hybridizations in carbon-adatom bonds, and the spin configurations. Whether there exist the diverse electronic properties (semiconductors or metals) and magnetic configurations (non-magnetism, ferromagnetism, or anti-ferromagnetism) is explored thoroughly. The theoretical predictions are compared with the up-to-date experimental measurements, especially for those from ARPES, STM, and STS. The complete and reliable results are very useful in the development of potential applications. Moreover, the first-principles cal-

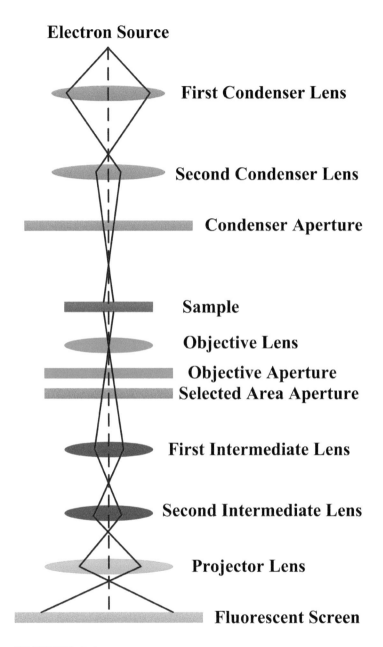

**FIGURE 2.1**
The TEM instrument in measuring electron diffraction pattern.

culations are available to determine the important hopping integrals (atomic interactions; the important parameters) in the tight-binding model [10].

The critical orbitals of carbons and adatoms are fully considered in the numerical calculations. The single- or multi-orbital hybridizations in chemical bondings, which dominate the feature-rich essential properties, are clearly identified from the atom dominance of energy band, the orbital-dependent charge distribution, and the orbital-projected DOSs. For pristine graphene nanoribbons, they can comprehend the dramatic changes of chemical bondings on the non-planar surfaces, e.g., the curvature effects related to the misorientation of $2p_z$ orbitals and the $sp^3$ hybridization of $(2s, 2p_x, 2p_y, 2p_z)$ orbitals [214]. Moreover, there exist the rich and unique orbital hybridizations in the various X-C and X-X bonds. They are responsible for the optimal geometric structure, the unusual electronic properties, and even diversify the magnetic configurations. As to the edge-carbon- and/or adatom-induced spin states, they are thoroughly examined from the magnetic moment, the atom-dominated spin degeneracy, the spatial spin distribution, and the spin-decomposed DOSs, being sensitive to the edge structures, the non-planar structures, the stacking configurations, the edge decoration, the kind of adsorption atatoms, the strength of X-C and X-X bonds, and the distribution and concentration of adatoms. In short, both orbital hybridizations and spin distributions will play critical roles in creating the diverse electronic properties and magnetic configurations, the FM/AFM/NM metals and semiconductors. The developed theoretical framework is very useful in the further studies on the emergent 1D and 2D materials, such as the essential properties of silicene-, germanene-, tinene-, and phosphorene-related systems.

## 2.2 Experimental instruments

Experimental measurements can directly identify the main characteristics of the essential properties; furthermore, the detailed comparisons with the theoretical predictions will provide the concise physical and chemical pictures to establish the basic concepts. Four main instruments throughout this work are TEM, STM, STS, and ARPES, in which the latter three also cover the spin-polarized measurements. They are very powerful in determining lattice symmetry, bond lengths, stacking configurations, spin distribution, local nanostructures, adatom positions, van Hove singularities (special structures) in DOS, spin-split states, energy dispersions of valence bands, band-edge states, and energy gap. A brief introduction to each characterization tool and the recent experimental progress is presented as follows.

Transmission electron microscopy is a microscopy technique in which an electron beam with a uniform current density is transmitted through an ultra thin specimen to form an image, owing to the interactions between incident charges and sample. The TEM was first demonstrated by Ernst Ruska and Max Knoll in 1931. Ruska was later awarded the Nobel Prize in physics for the development of TEM. TEMs can achieve a higher resolution than light microscopes, as a result of the smaller de Broglie wavelength of electrons. A TEM instrument, as shown in Figure 2.1, consists of an electron source, the first and second condenser lenses with aperture, specimen, an objective lens with aperture, intermediate and projector lenses, and a fluorescent screen. A set of condenser lenses helps to focus the beam on the sample. And then, an objective lens collects all the electrons after interactions, forms an image of the sample, and determines the limit of structure resolution. Moreover, a set of intermediate lenses magnifies this image and projects it on a fluorescent screen, a layer of photographic film, or a sensor. Compared to other microscopes, the main advantage of TEM measurements is to simultaneously provide accurate information in the real space (from the imaging mode) and reciprocal space (from the diffraction mode). The de Broglie wavelength of electrons of TEM is much smaller than atomic separations in the solids; therefore, it is achievable to observe crystal details below the atomic sizes. The resolution for TEM measurements is about 1-2 Å [183].

TEM is the powerful experimental technique for directly visualizing the crystal structure, locating and identifying the type of defects, and studying structural phase transitions. It has a very strong electrons' atomic scattering factor, being ~10,000 times that from the X-ray diffraction. This provides electron diffraction an advantage to detect even the weakest diffracted spot. However, the resolution of TEM is limited by spherical and chromatic aberrations of the lenses. More delicate techniques for improving the diffraction resolution become indispensable. By applying a monochromator and a Cs corrector into TEM, which is called high resolution TEM (HRTEM), the structural resolution can reach less than 0.5 Å [158]. HRTEM has been successfully and extensively used for analyzing crystal structures and lattice imperfections in various kinds of nanomaterials [286, 259]. The TEM/HRTEM measurements on graphene-related systems are very suitable in identifying the various nanoscaled geometric structures, such as the multi-walled cylindrical structures of carbon nanotubes [139, 166], the curved [166, 51, 171], folded [224, 387], and scrolled [345, 305, 311] profiles of graphene nanoribbons, as well as the stacking configurations and the interlayer distances of graphenes [45, 359, 181].

STM has become the most important experimental technique in resolving the surface structure since the first invention by Binnig and Rohrer in 1982 [29, 30]. This instrument is able to image the topographies of surfaces in real space with both lateral and vertical atomic resolution, such as the nanoscale bond lengths, crystal orientations, planar/non-planar structures, step edges, local vacancies, dislocations, adsorbed adatoms/molecules, and clusters/islands. The STM instrument, as shown Figure 2.2, consists of a conducting (semi-conducting) solid surface and a sharp metal tip, in which the distance of several angstroms is modulated by piezoelectric feedback devices.

A weak current is built by the quantum tunneling effect in the presence of a bias voltage between surface and tip. It has a strong relation with distance, being assumed to present the exponential decay form. This current flows from the occupied electronic states of tip into the unoccupied ones of surface under the positive bias voltage $V > 0$, or vice versa. In general, the quantum tunneling current is chosen to serve as a feedback signal. The most commonly used mode, a constant tunneling one (current and voltage), will be operated to resolve surface structure very accurately by applying a piezoelectric device. By combining a probe of metal tip and a precise scanning device, the full spectroscopic information of surface morphology is obtained at the preselected positions under the well-defined conditions. The structural response is hardly affected by the background effects, being attributes to an ultra-high vacuum environment. Up to now, the spatial resolution of STM measurements can reach 0.1 Å. In addition to STM, other experimental techniques are available in characterizing the geometric properties of low-dimensional systems, covering transmission electron microscopy (TEM) [325, 221], scanning transmission electron microscope (STEM) [45], atomic force microscopy (AFM) [237], and low-energy electron diffraction (LEED) [86]. It should be noticed that STM is very sensitive to electron spin by using a ferromagnetic/antiferromagnetic probe tip. This spin-dependent spectroscopic mode was first proposed by Pierce in 1988 and realized by Wiesendanger et al. in 1990 [273, 363]. The up-to-date spin-polarized STM (SP-STM) can provide detailed information of magnetic phenomena with atomic resolution, such as spin polarizations of individual adatoms on magnetic surfaces [379], and domain structures of magnetic elements [276, 296].

STM is a powerful experimental technique in revealing the spatially atomic distributions with the nanoscale precisions. The very accurate measurements on the carbon-related systems have confirmed the rich and unique geometric structures, illustrating the complex relations among the hexagonal lattice, the finite-size confinement, the flexible feature, and the diverse chemical bondings of carbon atoms. From the up-to-date STS observations, graphene nanoribbons present the nanoscale-width planar honeycomb lattice, accompanied with the non-chiral (armchair and zigzag) and chiral edge structures [294, 335, 233]. Furthermore, they could also be formed in the curved [335], folded [193], scrolled [345], and stacked lattice structures [148]. Carbon nanotubes possess the achiral and chiral arrangements of the hexagons on cylindrical surfaces [364, 256]. The atomic-scale measurements directly identify the AB, ABC, and AAB stacking configurations of few-layer graphenes [181, 293], the corrugated substrate and buffer graphene layer [64, 63], the rippled structures of graphene islands [241, 18, 87], and the adatom distributions on graphene surface [262, 19]. The layered graphite could exhibit the 2D networks of local defects on surface [53]; the pyridinic-nitrogen and graphitic-N structures [164]. In addition, the folded graphene nanoribbons are also confirmed by high-resolution TEM measurements [193].

STS, as shown in Figure 2.3, is an extension of STM by using more delicate

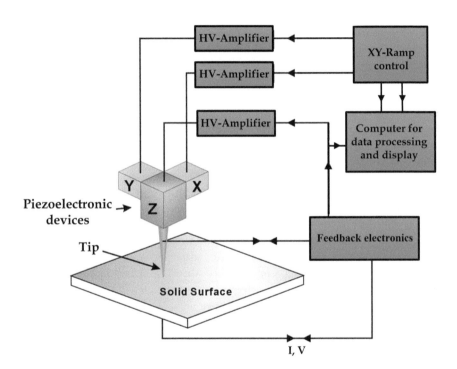

**FIGURE 2.2**
The schematic illustration of STM in observing surface configuration by controlling the tip height under a specific tunneling current.

operations and detailed analyses in the absence of feedback. This instrument is able to characterize the energy-dependent density of electronic states in condensed-matter systems, based on the relation between current and bias voltage. The I-V curves are obtained from a probing tip being fixed at the same height and scanning on a surface. They are closely related to the information of surface electronic states, when the tunneling current is assumed to be proportional to the constant states of a probing tip. The normalized differential conductance of the tip-surface tunneling junction is defined as (dI/dV)/(I/V), being interpreted as the DOS of surface structure. For example, as to 1D graphene nanoribbon/carbon nanotube, its relation with bias voltage corresponds to the energy dependence of DOS. In the STS measurements, the normalized differential conductance, being created by a small AC modulation of dV, can be examined by using a lock-in amplifier. Furthermore, the noise involved in the measured conductance will be greatly reduced. The up-to-date experiments can achieve the highest resolution of $\sim$ 10 pA. STS is available in the identification of the spin-split DOS, when the magnetic tips are utilized in the experimental observations. The spin-polarized STS (SP-STS) is very important in understanding the magneto-electronic properties, e.g., the metallic ferromagnetism and the semi-conducting anti-ferromagnetism [275]. Specially, STM and STS could be operated on the same sample to characterize the geometric and electronic properties simultaneously, such as the edge orientations and the width-dependent energy gaps of graphene nanoribbons [233, 364, 256, 134, 321, 70].

The STS measurements can efficiently examine the van Hove singularities due to the band-edge states and the semiconducting/semi-metallic/metallic behavior, since the tunneling conductance directly corresponds to DOS. They have successfully verified the diverse electronic properties in graphene nanoribbons [134, 321, 70], carbon nanotubes [364, 256], few-layer graphenes [231, 188, 72, 175, 378, 282, 274], adatom-adsorbed graphenes [64, 63, 114], and graphite [159, 187], as measured from the energy, number, intensity, and form of special structures in DOSs. As for graphene nanoribbons, the width- and edge-dominated energy gaps and the asymmetric peaks of 1D parabolic bands are identified from the precisely defined crystal structures [233, 134, 321, 70, 364, 256]. The similar strong peaks are revealed in carbon nanotubes, in which they present the chirality- and radius-dependent band gaps and energy spacings between two neighboring subbands [364, 256]. Specifically, armchair nanotubes belong to the 1D metallic systems with a finite DOS at the Fermi level. A lot of STS measurements conducted on few-layer and adatom-doped graphenes clearly show the low-lying DOS characteristics, covering a linear E-dependence vanishing at the Dirac point in a monolayer system [187], the asymmetry-created peak structures in bilayer graphene [231, 188, 72], an electric-field-induced gap in bilayer AB stacking and tri-layer ABC stacking [175, 378], a prominent peak at $E_F$ due to the partial flat bands in tri-layer and penta-layer ABC stacking [364, 256], a dip structure at $E_F$ accompanied with a pair of asymmetric peaks in tri-layer AAB stacking [364],

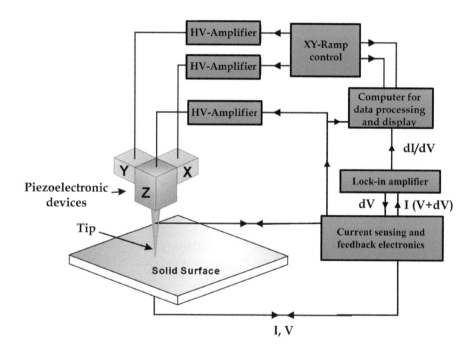

**FIGURE 2.3**
The schematic diagram of taking the current-dependent normalized conductance spectrum by using STS.

and a red shift of Dirac point arising from the metallic doping of Bi adatoms [64, 63]. The measured DOS of AB-stacked (Bernal) graphite is finite near the Fermi level characteristic of the semi-metallic property [187] and presents the splitting $\pi$ and $\pi^*$ strong peaks at deeper/higher energy [159].

Spin structures and their excitations, with the atomic resolution, play an important role in understanding magnetic properties and spin-dependent phenomena. The spin-polarized STM (SP-STM) technique is developed with a combination of the atomic-resolution capability of STM with spin sensitivity. SP-STM is utilized to take atomically resolved pictures of the magnetic ordering in a material by recording the constant current images which represent electron density near the sample surface, and simultaneously the spin polarization of charge carriers which is related to the local magnetic moment. It is very sensitive to the spin orientation of tunneling electrons. This magnetic imaging technique was first proposed by Pierce in 1988 [273] and realized by Wiesendanger et al. in 1990 [363]. The experimental setup for SP-STM measurement is shown in Figure 2.4. The essential fact is that the spin polarization of the tunneling current is a function of state energy. The spin polarization of electronic states that contribute to quantum tunneling depends on the sample bias voltage. For instance, when a finite negative sample bias (U) is applied between the tip and measured system, the occupied sample states in the range of U below the Fermi level will have contributions to the tunneling current. In the tunneling process, the electrons change into the unoccupied tip states with the range of U. The spin configuration of both tip and sample states dominate the quantum tunneling. In general, the spin polarization of the tunneling current varies with the bias voltage. Similarly for the case of SP-STS, the variations in the tunneling current are proportional to the differential conductance dI/dU.

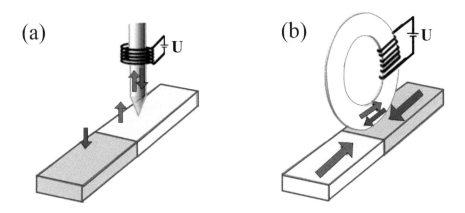

**FIGURE 2.4**
SP-STM experimental set-up: (a) the out-of-plane component, and (b) the in-plane component of the sample spin polarization.

Nowadays, the SP-STM and SP-STS provide the unprecedented insight about the diverse magnetic properties of the mainstream materials, leading to the discovery of new types of magnetic ordering under the nanoscale. SP-STM has been used to determine the magnetic domains and domain-wall structures at high spatial resolution of 10 nm, a length scale inaccessible by other magnetic imaging techniques [362]. The high-resolution SP-STM and SP-STS are also available in studying the spin-dependent scattering states around the individual impurity adatoms on magnetic substrates. For example, when O atoms are absorbed on Fe substrates, the spin-dependent scattering of Fe electrons with the $d$-like symmetry is revealed in real space by SP-STS [285]. The up-to-date SP-STM could provide the detailed information of magnetic phenomena with atomic resolution, such as spin polarizations of individual adatoms on magnetic surfaces [379], and domain structures of magnetic elements [276, 296].Moreover, the magnetic skyrmion and the ferromagnetic states have been observed by using SP-STM measurement for an epitaxial Fe film on Ir(111) substrate with the application of electric field [130]. Specifically, SP-STS is very important in understanding the magneto-electronic properties, e.g., the metallic ferromagnetism and the semiconducting anti-ferromagnetism [275]. The further SP-STM/SP-STS examinations are required to verify the rich and unique spin configurations in graphene-related1D and 2D systems.

ARPES is best for studying the quasi-particle energies of the occupied electronic states solids in the Brillouin zone [138], and such dispersion relations can directly examine the first-principles calculations. The ARPES chamber is accompanied with the instrument of sample synthesis to measure the *in situ* band structure. When a sample is illuminated by soft X-rays (Figure 2.5), the occupied valence electrons are excited to the unoccupied intermediate states after a dipole transition. Photoelectrons are stimulated by incident photons and escape outside of the material into the vacuum, and then they are counted by an angle-resolved energy analyzer. The total momenta of photoelectrons are evaluated from the free particle model, in which the parallel and perpendicular components are determined by the polar and azimuthal angles ($\theta$ and $\phi$ in Figure 2.4). The former are conserved during the photoemission process, while the conservation law is not suitable for the latter due to the breaking of translation symmetry along the normal direction. In general, ARPES measurements are mainly focused on two- and quasi-two-dimensional systems with the negligible energy dispersions perpendicular to surface. However, the non-conservation issue might be overcome by the important characteristic of the $k_z$-dependent band structure, as done for the 3D $\pi$ band of layered graphite using that at $k_z = 0$ [58]. Specifically, the ARPES measurements can provide energy widths of valence states, directly reflecting the many-body de-excitation effects due to the electron-electron and electron-phonon interactions [79]. Improvements in energy and momentum have become a critical factor for the investigation of low-dimensional materials. Up to now, the highest resolutions for energy and angular distributions are, respectively, $\sim$

1 meV and 0.1◦ in the UV regime [184]. Specifically, the spin-split energy bands could be examined by using the magnetic material as ARPES detector, e.g., the spin-resolved energy dispersions in topological insulators [281], ferromagnetic/anti-ferromagnetic materials [84], and superconductors [281].

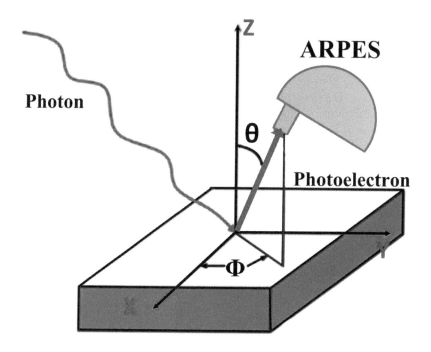

**FIGURE 2.5**
The photo-electronic spectrum of ARPES in determining the momentum-dependent valence state energies.

The high-resolution ARPES is the only experimental instrument able to directly measure the wave-vector dependence of the occupied energy bands. The measured results have verified the feature-rich band structures for the carbon-related systems, as observed under the various dimensions [257, 316, 38, 294, 327, 107], layer numbers [77, 257, 37, 329], stacking configurations [258], substrates [77, 257, 316], and adatom/molecule adsorptions [263, 398, 108]. Graphene nanoribbons are identified to possess 1D parabolic energy subbands centered at the high-symmetry point, accompanied with an energy gap and distinct energy spacings [294]. Recently, many ARPES measurements are conducted on few-layer graphenes, in which they have confirmed the linear Dirac cone in monolayer system [257, 316, 38], two pairs of parabolic bands in bilayer AB stacking [257, 258], the coexisting linear and parabolic dispersions in symmetry-broken bilayer system [37], the linear and parabolic bands in tri-layer ABA stacking [77, 257], the linear, partially flat and sombrero-shaped bands in tri-layer ABC stacking [77], the substrate-induced large energy spacing between the $\pi$ and $\pi^*$ bands in bilayer AB stacking [77, 257], the substrate-created oscillatory bands in few-layer ABC stacking [257]; the metal-semiconductor transition and the tunable low-lying energy bands after the molecule/adatom adsorptions on graphene surface [263, 398, 108], the AB-stacked graphite exhibits the 3D band structure [327, 107], with the bilayer- and monolayer-like energy dispersions at $k_z = 0$ and 1, respectively (K and H points in the 3D first Brillouin zone). Furthermore, there exists the strong trigonal warping effect around the KH axis, and a weakly dispersive band near the Fermi level attributed to the stepped surface with zigzag edge.

# 3

# Monolayer graphene nanoribbons

There are a lot of theoretical studies on pristine graphene nanoribbons without/with hydrogen passivation. In general, the essential properties are conducted by the tight-binding model [249, 323], the Hubbard model [121] with the AFM spin configurations and the first-principles method [324]. The first model [249] predicts that armchair systems are, respectively, metals and semiconductors for $N_A = 3I + 2$ and others under a uniform bond length/hopping integral, where $N_A$ is the number of dimer lines along the transverse direction. This corresponds to sample electronic states from those of monolayer graphene in the presence of open boundary conditions. In the further modification by including the nonuniform hopping integrals [323], all the systems belong to semiconductors. Energy gaps are predicted to be inversely proportional to nanoribbon widths, clearly indicating the quantum-confinement effect. Also, electronic properties are sensitive to the change in edge structure. Another archial systems [249], zigzag graphene nonoribbons, exhibits a pair of partially flat conduction and valence bands nearest to the Fermi level without energy gap and magnetism. They appear at large wave vectors covering the zone boundary; furthermore, the dispersionless range is about 1/3 of the first Brilloun zone for sufficiently wide systems ($N_Z \geq 30$; $N_z$ the number of zigzag lines). When the spin configurations are taken into consideration, there exists the AFM arrangement across the nanoribbon center by using the second and the third methods [121, 324]. All the zigzag graphene nanoribbons exhibit the semiconducting AFM, since the partially flat bands have an energy gap in the absence of spin splitting. Apparently, the spin-dependent electron-electron Coulomb intercations will affect the gound state energy, in which the spin arrangements are only created by the edge carbons. It should be noticed that their wave functions are mostly localized at the two zigzag edges. Specifically, the first-principles calculations could deal with the passivation of hydrogens at open boundaries. The effects on the fundamental properties are worthy of a systematic investigation,

On the experimental side, few- and multi-layer graphene nanoribbons could be produced by the various experimental methods [193, 294]. The achiral and chiral systems, which are, respectively, characterized by the zero and non-zero angle between the central line along $\hat{x}$ and the edge orientation in terms of the arrangement of carbon hexagons. They have been clearly identified from the high-resolution STM measurements [193], while the important differences in the boundary and central C-C bond lengths need to be measured in detail.

The former is very useful in the characterization of the essential properties in the individual systems. That is to say, the further experiments could directly examine the theoretical results. The ARPES measurements are successfully utilized to verify the low-lying valence bands in armchair graphene nanoribbons [294]. The experimental measurements on zigzag and chiral systems are absent up to now, i.e., the partially flat valence band belonging to the edge-localized states in the former requires the ARPES identifications. There exist plenty of STS examinations on the main features of DOSs covering energy gaps and van-Hove singularities [134, 321, 70]. They have confirmed the width-dependent energy gaps of the armchair and zigzag graphene nanoribbons, i.e., the inverse relation between energy gap and nanoribbon width is reliable under the theoretical predictions and the experimental examinations. However, it should further clarify the effects due to hydrogen passivations. Also, the square-root asymmetric peaks, which arise from the 1D parabolic dispersions, obviously appear in the DOS experiments. There are no verifications for the delta-function-like symmetric peaks and the plateau structures, being induced by the partially flat valence/conduction bands and the linear ones. In addition, the optical [112] and transport [397] experiments are useful in understanding band gaps and energy dispersions; furthermore, the magnetic measurements on pristine graphene nanoribbons could test the absence of magnetism.

The first-principles calculations within the density functional theory are useful in fully understanding the essential properties of pristine/H-passivated graphene nanoribbons. How to induce the drastic changes by the hydrogen termination is one of the focuses. The calculated results include the zero-temperature non-spin-polarized/spin-polarized ground state energies, the non-uniform bond lengths, the band structures, the energy gaps/spacing, the spatial charge densities, the spin arrangements near the zigzag edges, the magnetic moments, and the orbital-decomposed DOSs. The dependence on nanoribbon width, boundary structure, and hydrogen passivation will be investigated thoroughly. Obviously, the non-uniform bond lengths could exist in graphene nanoribbons systems without/with hydrogen terminations, mainly owing to the transverse quantum-confinement effects. They strongly depend on the positions of carbon atoms, especially for those near the two edges. The predicted results could be verified by the high-resolution STM measurements [294, 335, 233]. Energy bands might exhibit three kinds of dispersion relations, namely, the parabolic, linear, and partially flat ones. Some of them are closely related to the edge atoms and hydrogens. The up-to-date ARPES examinations could account for part of theoretical predictions. The width-dependent energy gaps/spacings are calculated for various armchair and zigzag graphene nanoribbons in the absence and presence of hydrogen passivations. They have been examined by the STS experiments under the hydrogen passivations in detail, and thorough comparisons are made. The band-edge states of band structure, which correspond to the van-Hove singularities in the wave-vector-energy space, display the special structures in DOS. In short, the chemical bondings and magnetic configurations, being, respectively, associated with the spatial charge and spin distributions, are proposed to explain the H- and edge-carbon-dominated energy bands and DOSs. The feature-rich graphene nanoribbons are expected to have high potentials in near-future applications, such as the electronic [117], spintronic [156], sensing [132], and energy [27] devices.

## 3.1 Armchair systems

The geometric and electronic properties of pristine and hydrogen-passivated armchair graphene nanoribbons (AGNRs) are investigated for various widths. The widths of AGNRs, as shown in Figures 3.1(a) and 3.1(b), are characterized by the number of dimers lines $N_A$ along $\hat{y}$, in which the periodical length in a unit cell along $\hat{x}$ is $I_c = 3b$ (b is the C-C bond length). For pristine systems, since the ribbon edge is not saturated with hydrogen atoms, the geometric reconstruction takes place, leading to the edge dangling bonds with the shorter bond lengths. For example, the $N_A = 10$ system has the outermost C-C length of 1.238 Å (Figure 3.1(a)), in which the C-C bonds are non-uniform, and their central lengths are close to those of monolayer graphene. Such a feature is hardly changed by the variation of ribbon width. The unstable dangling bonds provide suitable environment to form the C-H bondings. On the other hand, the hydrogen-terminated AGNRs exhibit the strong C-H $sp^2$ bonding with 1.09 Å bond length (discussed from charge distribution in Figure 3.5), which is slightly affected by different widths. Their edge C-C bonds are longer than those of pristine systems, e.g., 1.367 Å of the $N_A = 10$ system in Figure 3.1(b). Apparently, the boundary terminations have dramatically changed the bond lengths and thus the hopping integrals, especially for those near the two edges. However, they cannot destroy the $\pi$ and $\sigma$ bondings, respectively, due to the $2p_z$ and $(2p_x, 2p_y)$ orbitals; that is, they do not affect the orbital dominance for the low- and high-energy (deep-energy) electronic states.

Electronic properties of pristine AGNRs are mainly determined by the edge structure, dangling bonds, and quantum confinement effects. The unoccupied conduction bands are asymmetric to the occupied valence bands about the Fermi level ($E_F = 0$), as shown in Figure 3.2(a) for the $N_A = 8$ armchair system. Most of energy bands exhibit the strong parabolic dispersions that belong to the extended states with wide spatial distributions. There exist four energy bands (large circles) with localized carrier distributions near the ribbon edges, in which the radius of each circle represents the contribution percentage from the edge atoms. They principally come from the edge dangling bonds. These edge-atom-dominated conduction and valence bands are doubly degenerate near $k_x = 1$; furthermore, the valence ones possess weak energy dispersions at about $-2.5$ eV $\leq E^v \leq -2.2$ eV. The ARPES measurements on them are absent up to now. They might be associated with the specific energy

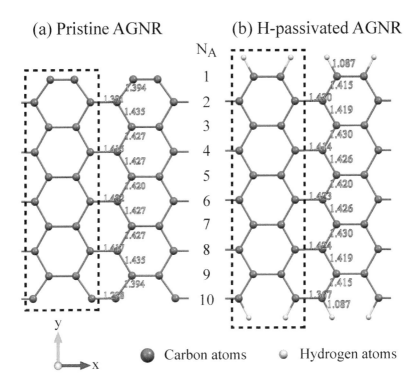

**FIGURE 3.1**
The geometric structure of $N_A = 10$ armchair graphene nanoribbons (a) without and (b) with hydrogen passivations in the presence of the non-uniform bond lengths.

gaps. For $N_A = 3I - 1$ and $3I + 1$ armchair graphene nanoribbons, the edge-atom-induced conduction ones at $k_x = 1$ and the first valence band (counted from $E_F$) at $k_x = 0$, respectively, correspond to the lowest unoccupied state (LUS) and the highest occupied state (HOS). These two kinds of armchair systems belong to indirect-gap semiconductors, e.g., $N_A = 8, 10, 11; 13$ in Figures 3.2(a), 3.2(c), 3.2(d), and 3.2(f), respectively. On the other hand, the $N_A = 3I$ systems, as indicated in Figures 3.2(b) and 3.2(e), possess the LUS and HOS in the first pair of conduction and valence bands at $k_x = 0$, respectively; that is, they are direct-gap semiconductors. The calculated results are consistent with the previous calculations [323]. Specifically, indirect and direct energy gaps could be examined from the experimental measurements on optical spectra [112], and transport properties [397].

The hydrogen edge terminations dramatically alter the electronic properties of armchair graphene nanoribbons, as shown in Figures 3.3(a)–3.3(f). The edge-atom-created conduction bands and weakly dispersive valence band vanish for various H-passivated armchair systems, indicating the absence of dangling bonds. The strong C-H bondings and the very low bound state energy of H-1s orbital are responsible for the drastic changes in energy bands. The edge C and H atoms make important contributions to certain valence bands in the range of $-6$ eV $\leq E^v \leq -3$ eV. The C-H-dominated energy band width is wider than 2.5 eV. This further requires the high-resolution ARPES verifications. The orbital hybridizations in C-C bonds will be analyzed using the orbital-decomposed DOSs. The LUS and HOS, respectively, correspond to the first conduction and the first valence band (counted from $E_F$) at $k_x = 0$. Hence, all the H-passivated AGNRs belong to semiconductors with direct band gaps. In general, the predicted energy gaps could be examined by the experimental measurements on band structure, DOS, optical spectra, and transport conductance. The energy gaps are sensitive to the ribbon width, as indicated by Figures 3.4(a) and 3.4(b). The dependence of $E_g$ on $N_A$ can be sorted into three groups, $N_A = 3n$, $3n + 1$, and $3n + 2$. The largest energy gap occurs for $N_A = 3n + 1$, and the smallest one for $N_A = 3n$ at the same $n$ value. Both pristine and H-passivated AGNRs present the similar features in energy gaps.

In general, energy gaps, as clearly indicated in Figures 3.4(a) and 3.4(b), strongly depend on nanoribbon width. They could be classified into $N_A = 3I_1$, $3I$ & $3I + 1$, being regardless of hydrogen passivations. Band gaps decline quickly in the increase of $N_A$; furthermore, there exist inverse relations from the well fittings. Among three categories, the $N_A = 3I + 1$ armchair systems exhibit the maximum energy gaps, while the pristine $N_A = 3I$ ones (the H-terminated possess $N_A = 3I - 1$ ones) the minimum energy gaps. Specifically, the hydrogen terminations could enhance/reduce energy gaps for $N_A = 3I - 1$ & $3I + 1/3I$. Only the pristine $3I \pm 1$ systems have indirect energy gaps.

The modifications on the tight-binding model are very useful to understand the semiconducting behavior of planar graphene nanoribbons with hydrogen passivation, the strong dependence of energy gap on the edge structure and

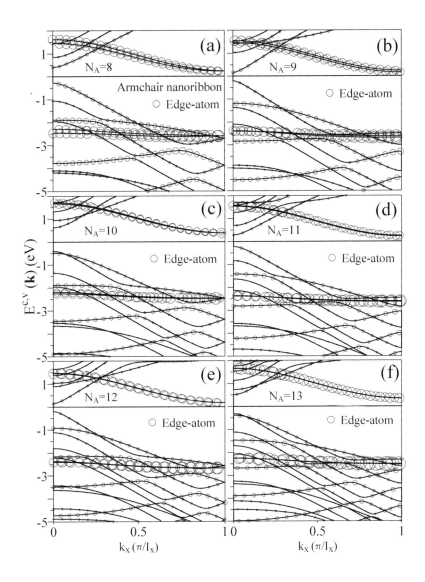

**FIGURE 3.2**
Energy bands of armchair graphene nanoribbons for (a) $N_A$=8, (b) 9, (c) 10, (d) 11, (e) 12, and (f) 13 in the absence of hydrogen passivation.

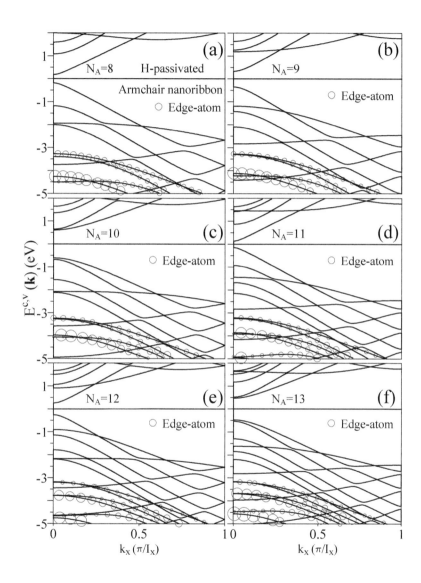

**FIGURE 3.3**
Similar plot as Figure 3.2, but shown with hydrogen passivation.

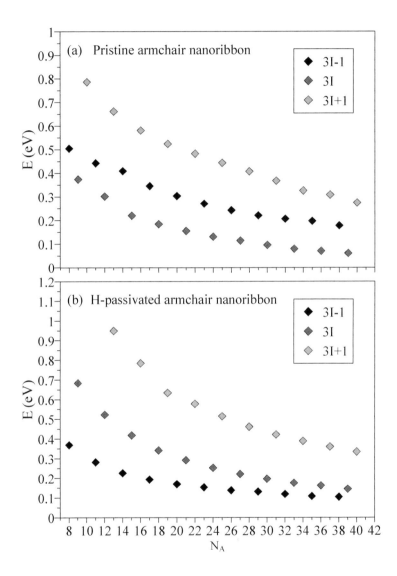

**FIGURE 3.4**

The width-dependent energy gaps of armchair graphene nanoribbons (a) without and (b) with hydrogen termination for the $N_A = 3I - 1$, $3I$, and $3I + 1$ categories.

width. When only the uniform $\pi$ bondings (the same nearest-neighbor hopping integral of $2p_z$ orbitals) is taken into account [102, 249, 101, 349], $N_A = 3I$ and $3I+1$ armchair nanoribbons are semiconductors with band gaps inversely proportional to width, being consistent with those from the first-principles method. The critical factor for moderate energy gaps is the open boundary condition.

On the other hand, $N_A = 3I + 2$ armchair nanoribbons and all zigzag systems are predicted to be gapless, since they, respectively, possess a pair of linear bands and partially flat bands at the Fermi level. This is in sharp contrast with the first-principles calculations. The obvious discrepancy could be solved by considering the non-uniform atomic interactions and the spin configuration (Hubbard model) near the armchair and zigzag edges, respectively [323, 121, 289, 96]. The chemical bondings between the passivated carbon and hydrogen atoms could induce the non-uniform C-C bond lengths and thus the position-dependent hopping integrals [323]. As a result, such armchair systems present small gaps due to the separated parabolic valence and conduction bands; that is, their electronic states cannot sample the Dirac point of monolayer graphene. According to the Hubbard Hamiltonian with the spin-related on-site Coulomb interactions [121, 289, 96], the anti-ferromagnetic and ferromagnetic spin configurations, which are, respectively, distributed on the distinct- and same-side zigzag edges, correspond to the ground state and are responsible for the separation of the edge-localized valence and conduction bands.

The carrier density ($\rho$) and the variation of carrier density ($\Delta\rho$) can provide useful information on the orbital bondings and energy bands. The former directly reveals the bonding strength of C-C bonds near the center or edge of graphene nanoribbon, as illustrated in Figures 3.5(a) and 3.5(b). All the C-C bonds possess the strong covalent $\sigma$ bonds and the relatively weak $\pi$ bonds simultaneously (black and gray rectangles). Such bondings are enhanced for the edge C-C bonds in pristine systems, as shown by the dark red region in Figure 3.5(a). This is responsible for the shortened bond lengths and the energy range of edge-atom-dominated bands. As for H-terminated ones, the C-H bondings are revealed by the red region between them, while the edge C-C bonds are weakened under the loss of the $\sigma$ electrons (Figure 3.5(b)). These features reflect the variation in bond lengths and can be further analyzed by the variation of carrier density. $\Delta\rho$ is obtained by subtracting the carrier density of an isolated carbon from that of an armchair graphene nanoribbon. Carbon atoms contribute four valence electrons to create two specific chemical bondings, namely the strong $\sigma$ bonding of $(2s, 2p_x, 2p_y)$ orbitals and the $\pi$ bonding of $2p_z$ orbitals. The former induce the increased charge density between two carbon atoms (red parts in Figure 3.5(c)). There are more charges distributed at the ribbon edge (dark red), indicating the strength of the C-C triple bond. $\Delta\rho$ in H-passivated AGNRs present a clear image of charge transfer from edge carbons to hydrogens (red parts near H atoms in Figure 3.5(d)). This accounts for the vanishing edge-atom-created energy bands.

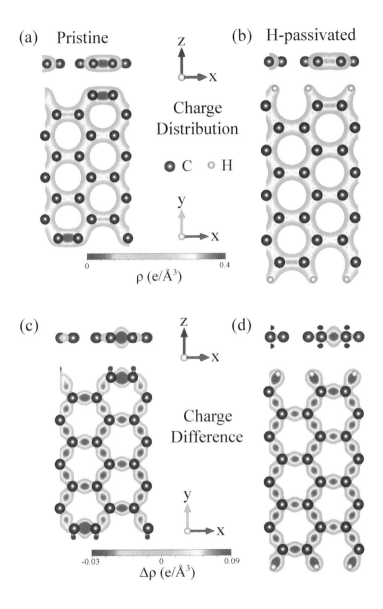

**FIGURE 3.5**
The spatial charge distributions of $N_A = 10$ armchair system in the (a)/(b) absence/presence of hydrogen terminations for (c)/(d) the total charge/charge difference.

DOSs directly reflect the main features band structures, in which the orbital-decomposed ones can understand the orbital hybridizations in C-C and C-H bonds. There are a lot of special structures in DOSs due to the band-edge states. Most of van Hove singularities, which come from parabolic energy dispersions, appear in the square-root divergent form because of the 1D characteristics, as those indicated in Figures 3.6(a)–3.6(h). Their heights are inversely proportional to the band curvatures. The number of asymmetric prominent peaks increases as the nanoribbon width grows. A pair of asymmetric peaks, which is centered at the Fermi level, could characterize an energy gap. Such structures might not be available in distinguishing the direct- or indirect-gap characteristics, since the low-lying double peaks are almost absent in the conduction band owing to the broadening effect (Figures 3.6(a)–3.6(d)). That is to say, the STS measurements could not be utilized to identify the direct or indirect gaps. Specifically, the edge carbon atoms make important contributions to the first conduction-band peak nearest to $E_F$. Few of special structures exhibit in the delta-function-like form corresponding to the partially flat edge-atom-dominated valence energy bands in pristine armchair systems, e.g., DOSs in the range of $-2.5$ eV $\leq E^v \leq -2.2$ eV in Figures 3.6(a)–3.6(d). They become asymmetric prominent peaks at deeper energies, as indicated in Figures 3.6(e)–3.6(h) below $E < -3$ eV by the light-blue peaks and those accompanied with them.

The orbital-projected DOSs show that the main contributions are due to the outer four orbitals of carbon and 1s orbital of hydrogen. For pristine armchair systems, the $\pi$ and $\sigma$ pronounced peaks, respectively, occur at the valence range of $E < 0$ and $E < -3$ eV, while both of them are revealed at the whole conduction range, mainly owing to the edge carbons with the dangling bonds. This indicates the existence of the partial $sp^3$ near the armchair boundaries. The low-lying conduction structures are only dominated by the $2p_z$ orbitals after the hydrogen terminations (Figures 3.6(e)-3.6(h)); that is, the $\pi$ bonding determines the low-energy essential properties of the H-passviated armchair graphene nanoribbons. Furthermore, the $1s$- and $(2p_x, 2p_y)$-related asymmetric peaks happen simultaneously, revealing the strong orbital hybridizations among three orbitals.

## 3.2   Zigzag systems

Zigzag graphene nanoribbons exhibit the similar geometric properties and thus the chemical bondings near the two boundaries, as observed in armchair ones. For example, the edge C-C bond lengths of $N_z = 10$ zigzag systems are, respectively, 1.382 Å and 1.407 Å in the absence and presence of hydrogen passivations (Figures 3.7(a) and 3.7(b)). The former is longer than that (1.238 Å) in armchair systems, so that the zigzag edge reconstruction is less serious. The

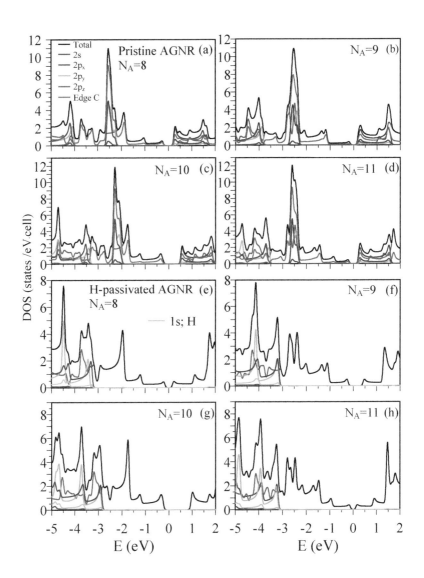

**FIGURE 3.6**

The orbital-decomposed density of states for (a) $N_A = 8$, (b) 9, (c) 10, and (d) 11 armchair graphene nanoribbons, and (e) $N_A = 8$, (f) 9, (g) 10, and (h) 11 systems with hydrogen teminations.

C-H bond length ($\sim$ 1.09 Å) hardly depends on the edge structures (Figures 3.7(b) and 3.1(b)), indicating the identical orbital hybridizations. The non-uniform geometric structure (hopping integrals) occurs on the same $(x, y)$ plane. It cannot fully change the $\pi$ bonding, but induce the significant distortion, especially for the existence of the edge-localized electronic states. On the other hand, the most important difference between zigzag and armchair systems is that the former possess the spin-up and spin-down magnetic configuration on the two distinct edges. The ground-state energy differences in the absence and presence of spin distributions are very obvious, in which they are, respectively, more than 0.8 eV and 0.2 eV for the pristine and H-terminated zigzag systems (the red and black diamonds in Figure 3.7(c)). Specially, their dependence on nanoribbon widths is rather weak, directly reflecting the dominance of edge carbons in the fundamental properties. The STM measurements have been successfully utilized to identify the zigzag and armchair edges and even the chiral structures [335]. They are useful in verifying the C-C and C-H bond lengths.

Zigzag systems belong to semiconductors in the presence of edge spin configurations. The non-spin-polarized arrangement has the higher ground state energy except for very wide zigzag systems. The spin-polarized configuration is used in the calculations. For example, the energy difference per carbon atom between these two kinds of spin configurations is 0.846 eV for $N_z = 10$. Their energy bands have strong dispersion relations, as shown in Figures 3.8(a)–3.8(f) for various ribbon widths. The occupied valence bands are asymmetric to the unoccupied conduction bands about $E_F = 0$. The low-lying valence and conduction bands exhibit a special valley centered at $k_x = 2/3$. This further illustrates that the low-energy electronic states of graphene nanoribbons are sampled from those near the Dirac points of monolayer graphene (the K and K' valleys). A pair of partially flat bands at $k_x \geq 2/3$ are revealed within $-1.6$ eV$< E^{c,v} < 1.0$ eV. They have been examined to originate from edge carbons, as indicated by the blue circles near the zone boundary ($k_x = 1$). The HOS and LUS come to exist at about $k_x = 2/3$, which determines the direct energy gaps for zigzag systems. Such band gaps become smaller when the ribbon width increases. For the wider zigzag nanoribbons, the more energy subbands at deeper/higher energy ($|E^{c,v}| > 2$ eV). Specially, there exists the edge-atom-induced valence and conduction bands. The former has the band width about 1.2 eV, and the latter exhibits the very weak dispersion centered at 1.0 eV. These two kinds of energy bands vanish after hydrogen terminations, as shown in Figures 3.9(a)–3.9(f). The edge carbon atoms and the H atoms contribute to certain energy bands deeper than $-3$ Ev, being similar to those of H-passivated armchair ribbons. All the H-passivated zigzag systems present smaller energy gaps, compared with pristine ones. However, $E_g$ is also determined by the $k_x = 2/3$ states of the partially flat bands. This clearly illustrates the robustness of the edge-localized states, independent of the hydrogen terminations. Specifically, the electronic states are doubly degenerate for the spin degrees of freedom in the presence/absence of hydrogen passi-

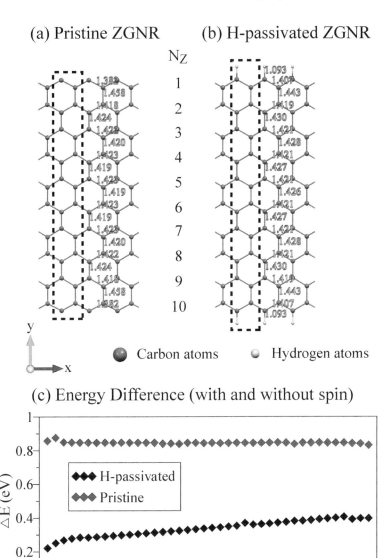

**FIGURE 3.7**
The $x - y$-plane optimal geometric structure of $N_Z = 10$ zigzag nanoribbons (a) without and (b) with hydrogen terminations under the non-uniform bond lengths. (c) clearly shows the ground-state energy differences between without and with zigzag edge spin distributions.

vations, especially for the partially flat valence and conduction bands. They correspond to the AFM configuration across the ribbon center and the FM one on the same side (discussed later in Figure 3.11). The net magnetic moment is zero under the AFM arrangement in any zigzag systems. The magnetic properties are also consistent with the theoretical predictions of the Hubbard model based on a bipartite lattices [323].

Energy gap at $k_x = 2/3$ and energy spacing ($E_s$) at $k_x = 1$, which are associated with the partially flat valence and conduction bands, are worthy of a closer examination. Apparently, they, as indicated in Figures 3.10(a)–3.10(b), decline quickly in the increment of $N_z$, mainly owing to the reduced finite-size effect. The quantum confinement and the significant dangling bonds are responsible for the width-dependent energy gaps without hydrogen passivations (the black diamonds in Figure 3.10(a)). The hydrogen adsorbates could greatly suppress the edge dangling C-C bonds and thus reduce band gaps, especially from the narrow systems (the red diamonds). Specifically, energy gaps arise from the magnetic ordering, since the non-spin-polarized arrangement in H-passivated zigzag graphene nanoribbons only creates the merged pair or the gapless behavior. The first-principles calculations agree with the Hubbard model, accompanied with the AFM configuration. The theoretical predictions on band gaps of H-terminated zigzag systems have been verified by the STS measurements [134, 321, 70]. Energy spacing at the zone boundaries is very important in the characterization of band structure; furthermore, the corresponding band-edge conduction state is very close to the edge-atom-dominated energy band (Figures 3.8(a)–3.8(f)). They will be revealed as a very strong double-peak conduction structure in DOSs (discussed later in Figures 3.12(a)–3.12(d)). They are very easy to identify these two van Hove singularities. $E_s$ is weakly dependent on the nanoribbon width, regardless of hydrogen passivation (the black and red diamonds in Figure 3.10(b)).

The orbital bondings and related energy bands could be understood from the carrier density and the variation of carrier density. The former directly illustrates the stronger bonding strength of the edge C-C bonds, as obviously indicated in Figure 3.11(a) for $N_z = 10$ zigzag graphene nanoribbon. Such enhanced bondings are similar to those in armchair systems, clearly revealing the shortened edge C-C bonds. The H-terminated system has the very strong C-H bondings shown by the red region between H and C atoms, in which the edge C-C bonds are slightly weakened (Figure 3.11(b)). There is no evidence related to the $\hat{z}$ direction; that is, the $2p_z$ orbitals of carbons do not participate in the orbital hybridizations in the C-H bonds. The electron transfer during the self-consistent calculations is presented by $\Delta\rho$ in Figures 3.11(c) and 3.11(d). Electrons in a pristine system are transferred to the edge zigzag chains (the dark red regions). $\Delta\rho$ in H-passivated ZGNRs exhibit significant charge transfers to hydrogens (red region near H atoms). This is the main reason for the vanshing edge-atom-induced energy bands. Specifically, the anti-ferromagnetic configuration across the nanoribbon center/the ferromagnetic one on the same side is shown by the spatial distribution of the charge difference between two

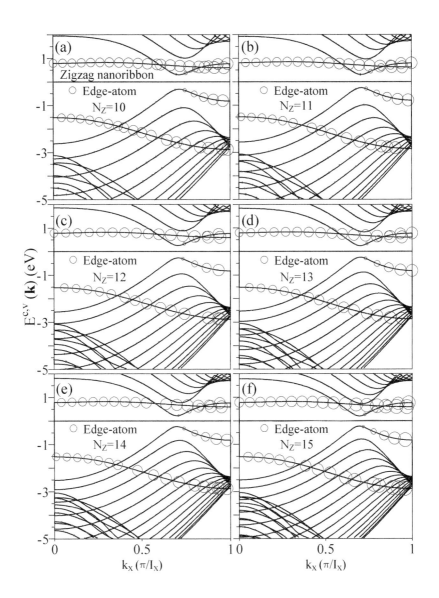

**FIGURE 3.8**
Electronic structures for zigzag graphene nanoribbons with (a) $N_Z=10$, (b) 11, (c) 12, (d) 13, (e) 14, and (f) 15 in the absence of hydrogen passivation.

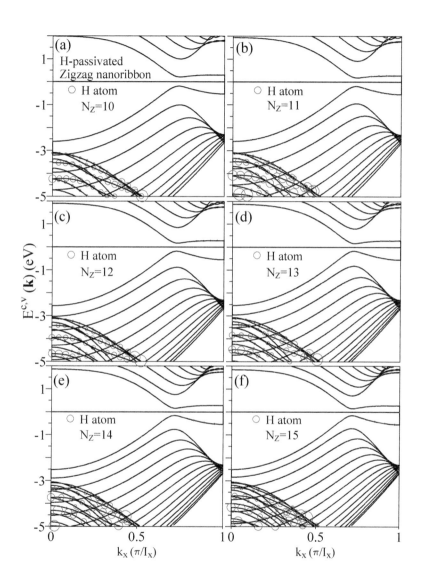

**FIGURE 3.9**
Similar plot as Figure 3.3, but shown with hydrogen passivation.

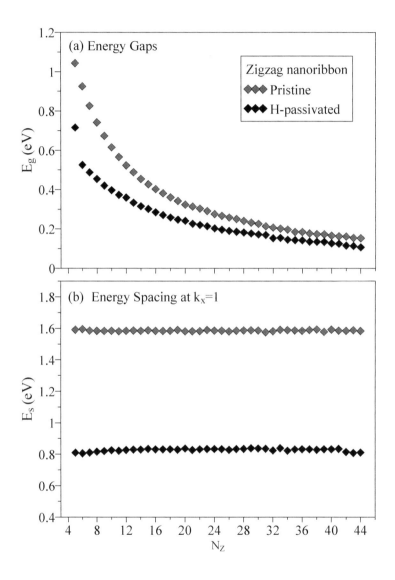

**FIGURE 3.10**
The width-dependent energy gap (a) and energy spacing at zone boundary (b) due to a pair of partially flat valence and conduction bands for zigzag graphene nanoribbons without and with hydrogen passivations.

spin states in Figures 3.11(e) and 3.11(f). The narrow red and blue regions near the two zigzag edges indicate the fact that the hydrogen terminations can effectively suppress the edge magnetic moments. This AFM has the zero net magnetic moment, since two edges in a zigzag system are symmetric about the nanoribbon center.

DOSs directly reflect the special features of band structures in zigzag graphene nanoribbons, clearly shown in Figures 3.12(a)–3.12(h). For the $N_Z = 10$ system, they exhibit a lot of asymmetric peaks and a few symmetric ones due to the parabolic and partially flat energy bands, respectively. A pair of asymmetric peaks in the absence of hydrogen terminations (Figures 3.12(a)–3.12(d)), which is centered about the Fermi level, exactly defines the energy gap. They arise from the valence and conduction valleys nearest to $E_F$ (Figures 3.8(a)–3.8(f)); furthermore, they belong to part of the edge-localized bands. Such structures are associated with the $\pi$ bonding of $2p_z$ orbitals extended in the plane (the red curves), but not dominated by the edge atoms (the brown curves). The most contributions from the edge atoms appear at the low energies, covering the asymmetric peak at $E \sim -0.8$ eV and the prominent double-peak structure at $0.6$ eV$\leq E \leq 0.9$ eV. The latter is created by two special energy bands, namely, the edge-atom-dominated conduction band with the weak dispersion and the edge-localized parabolic band initiated from the zero boundary. The double-peak structure is mainly determined by the dangling bonds of the edge atoms. The orbital hybridizations in this bond are further identified to be the $sp^3$ bondings, since the four $(2s, 2p_x, 2p_y, 2p_z)$ orbitals exhibit obvious peaks simultaneously (the purple, blue, green, and red curves). In addition to the $\pi$ bondings, the $\sigma$ bondings also make important contributions to the valence DOS, especially for the deep-energy one. Their contributions are initiated from $-1.5$ eV ($2p_x$ orbital corresponding to the blue curve), being in sharp contrast with that (lower than $-3$ eV) of layered graphene [340]. As for the special structures induced by the dangling bonds, they are vanishing after the hydrogen passivations, as indicated in Figures 3.12(e)–3.12(h). Specifically, there are one strong conduction peak and two valence peaks near the Fermi level arising from the pair of edge-localized energy bands, mainly owing to the almost dispersive conduction band in the range of $K_x \geq 2/3$ (Figures 3.9(a)–3.9(f)). Moreover, the C and H contributions could occur at deep energy lower than $-3$ eV, directly revealing the rather strong C-H bonds. The $(2s, 2p_x, 2p_y)$ orbitals, as illustrated by the orbital-decomposed DOSs, take part in the C-H bonds. That is to say, there exist the $sp^2s$ bondings. In short, zigzag and armchair graphene nanoribbons are very different from each other in the low-lying special peaks because of the absence/presence of the edge-localized energy bands and the edge-atom-dominated conduction band in the latter/former. It should be noticed that the latter have a special double-peak structure at $-2.5$ eV$\leq E \leq -2.2$ eV.

One of the most important electronic properties, energy gaps, in the absence of external fields and mechanical strains, strongly depends on edge structures and nanoribbon widths. They could be directly examined from

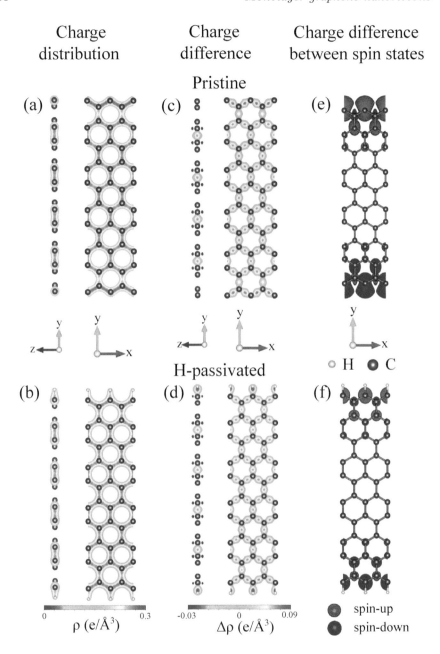

**FIGURE 3.11**
The spatial charge distributions of $N_z = 10$ zizag system in the (a)/(b) absence/presence of hydrogen terminations for (c)/(d) the total charge/charge difference. Also shown in (e) and (f) are the spin charge difference between spin states.

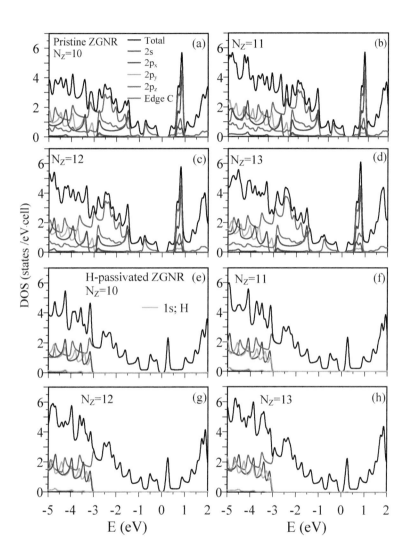

**FIGURE 3.12**

The orbital-projected DOS for (a) $N_z = 10$, (b) 11, (c) 12, and (d) 13 zigzag graphene nanoribbons, and (e) $N_z = 10$, (f) 11, (g) 12, and (h) 13 systems in the presence of hydrogen teminations.

STS, transport, and optical measurements, especially for the first method belonging to the most efficient tool. Under the hydrogen passivations, the STS experiments have confirmed three kinds of energy gaps in armchair graphene nanoribbons (Figures 3.4(a) and 3.4(b)) and the edge-localized-state gaps in zigzag systems (Figures 3.10(a) and 3.10(b)), in which they decline quickly in the increase of width [193]. Only a few STS measurements are conducted on graphene nanoribbons with/without hydrogen terminations, e.g., $N_A = 7$ armchair graphene nanoribbon [161, 294, 134]. To identify the effects due to the dangling bonds at two edges (Figures 3.4(a) and 3.10(a)), a systematic study on them is required by using STS, covering the differences between $N_A = 3I$ and $3I - 1$ and the reduced energy gap for any zigzag nanoribbon. The magnitude of $E_g$ will determine the temperature-dominated conductance, so it could be estimated from the measured transport property. The transport measurements are utilized to verify the specific relation that energy gaps are inversely proportional to nanoribbon widths for any graphene nanoribbons [397]. In addition, they also examine the chiral-angle-dependent energy gaps. The optical instruments, as done for carbon nanotubes [386, 7], are suitable to examine the direct-gap systems. However, there are no optical measurements up to now.

The ARPES measurements have confirmed parabolic valence bands, energy spacings, and band gaps in armchair graphene nanoribbons in the presence/absence of hydrogen terminations, such as $N_A = 7$ [294, 321, 318] and 9 systems [308, 331]. Specifically, ARPES is conducted on an $N_A = 7$ armchair system without hydrogen passviation [321], but on Au(111) in which the measured band width is about 3 eV below the Fermi level. Such measurement does not verify the partially flat valence bands at $-2.5$ eV$\leq E^v \leq -2.2$ eV due to the significant dangling bonds, since it might have the orbital hybridizations in C-Au bonds. The energy range needs to be extended to identify the deeper $(C, H)$-co-dominated energy bands, since the deeper-energy experimental measurements are absent up to now. On the other hand, there are no ARPES experiments on the significant effects due to the hydrogen terminations, edge structures, and nanoribbon widths, such as the existence of the edge-carbon-dominated valence bands, the (C,H)-co-induced energy bands, the edge-localized states in zigzag systems, the inverse relations between energy gaps and widths, and the H-passivation-reduced/enhanced band gaps. Such verifications could promote the thorough understanding of the critical chemical bondings, covering the dominating $\pi$ and $\sigma$ bondings for the C-C bonds, the partial $sp^3$ hybridizations in the edge dangling bonds, and the strong $sp^2s$ hybridizations in the C-H bonds.

The essential properties of graphene nanoribbons are very sensitive to the external fields [56] and the mechanical strains [55]. The transverse electric field and the uniform deformation could induce the semiconductor-metal transitions. The former creates the Coulomb potential energies on the different sites, and the latter leads to the variations of the hopping integrals along the distinct C-C bonds. Both of them also strongly depend on the edge structures and the nanoribbon widths. They are very efficient methods in tuning the electronic properties; that is, they could be utilized to modulate energy gaps. The relation between the energy gap and electric field strength is monotonous for sufficiently wide systems, while the dependence of $E_g$ on the mechanical strain might become oscillatory under the uniaxial stress. On the other hand, they cannot destroy the state degeneracy and the magnetic properties, covering the subband degeneracy in any graphene nanoribbons and the AFM configurations in zigzag ones.

When graphene nanoribbons exist in a uniform perpendicular magnetic field $(B_z\hat{z})$, the fundamental properties are dramatically changed during the variation of $B_z$. Based on the generalized tight-binding model [136, 203, 75], the vector potential associated with $B_z$ will modify the neighboring hopping integrals by the Peierls phases. There exist plenty of Landau subbands under the magnetic length $(\sqrt{1/B_z}?)$ longer than the nanoribbon width [136, 203, 75]. They do not belong to the Landau levels with the fully magnetic quantization, mainly owing to the finite-size effect. In addition, it is very frequent to observe the Landau levels in 2D few-layer graphenes and emergent materials [391]. The dispersionless ranges (the highly degenerate states) are, respectively, centered about $k_x = 0$ and $\pm 2/3$ for armchair and zigzag systems; that is, each Landau subband consists of the dispersionless and parabolic form. The low-lying Landau subbands are relatively easily created from the electronic states close to the Fermi level, since the magnetic field will compete with the kinetic energy (from the hopping integrals). The $N$-layer systems possess $N$ pairs of valence and conduction Landau subbands; furthermore, their magnetic wave functions are similar to each other. The quantum number $(n^v/n^c)$ of each Landau subband is characterized by the number of zero points in the oscillatory probability distribution of the dominating subenvelope function. The layered systems have the asymmetric Landau subband spectra about the Fermi level because of the interlayer atomic interactions except for monolayer one. In general, the magnetic energy spectra might exhibit the non-crossing, crossing, and anti-crossing behaviors, depending on the number of layers and stacking configurations. It should be noticed that the semiconductor-metal transitions cannot be driven by the magnetic field. The main features of magneto-electronic properties, the $B_z$-dependent wave functions and energy spectra are directly reflected in the optical absorptions, especially for the specific magneto-optical selection rules (discussed below). Specifically, the magnetic unit cell is too large to deal with using the first-principles calculations.

The occupied electronic states are excited to the unoccupied ones, while graphene nanoribbons are present in an electromagnetic field. For 1D systems, the optical absorption spectra might exhibit the square-root asymmetry peaks, the delta-function-like symmetric peaks and the plateaus, respectively, corresponding to the parabolic, partially flat, and linear energy bands. They are mainly determined by the joint density of states and the electric dipole moment. Among all the absorption structures, most of them belong to the first type, and few of them are the second and third types. According to the theoretical predictions of the tight-binding model, there exist the edge-dependent selection rules in the absence of external fields, namely, $\Delta J = J^v - J^c = 0$ and $|\Delta J| = 2I + 1$ for armchair and zigzag graphene nanoribbons, respectively. $J^{c,v}$ is the subband index counted from the Fermi level, and it is not a good quantum because of the open boundary condition. These rules could be understood from the specific relations among the valence and conduction subbands, regardless of the numerical techniques. It is worthy of completely experimental measurements on optical absorption spectra, especially for the verifications of the edge-dominated selection rules. The well-behaved standing waves are dramatically changed into the localized Landau distributions with the dominating oscillatory modes, if a uniform perpendicular magnetic field is applied to a graphene nanoribbon. As a result, the inter-Landau-subband optical excitations at low energy agree with the magneto-optical selection rule of $|\Delta n| = 1$, in which the absorption peaks exhibit the delta-function-like form. The higher-energy absorption structures arise from the edge-dependent selection rule; furthermore, they belong to the asymmetric peaks. That is to say, the magneto-optical absorption spectra possess the critical characteristics of the geometry and magnetic quantization simultaneously.

# 4

## Curved and zipped graphene nanoribbons

A curved graphene ribbon corresponds to an unzipped carbon nanotube. The systematic studies have been made for the latter since the successful synthesis using the arc-discharge evaporation in 1991 [139], such as synthesis [337, 220, 83, 288, 95], characterization [364, 256, 20, 283, 25, 78], property [153, 149, 43, 272, 162, 287], and application [42, 334, 235, 312, 248]. The former could be produced from the latter by the various physical and chemical processes [166, 51, 171, 147, 146, 93, 264, 332, 261, 155, 47, 167, 310, 145, 369], as discussed in the introduction (page 2). The previous discussions clearly show that all the planar graphene nanoribbons exhibit the semiconducting behavior. On the other hand, the cylindrical carbon nanotubes are metals or direct-gap semiconductors sensitive to the chirality and radius. Metallic nanotubes are exclusively comprised of either armchair nanotubes or very small zigzag nanotubes with radii $< \sqrt{3}$b (b is the C-C bond length) [245]. The theoretical calculations [150, 314], and experimental measurements [260] have confirmed the curvature effects, the misorientations of $2p_z$ orbitals and hybridizations of carbon four $(2s, 2p_x, 2p_y, 2p_z)$ orbitals, on a cylindrical surface, leading to the geometry-dependent energy gaps. Apparently, for the theoretical calculations, the essential properties are expected to present the dramatic transformation during the unzipping process. For example, the quantized quasi-Landau levels and the magneto-optical selection rule could survive in curved graphene nanoribbons, while they are almost absent in cylindrical carbon nanotubes [75, 3, 6].

The essential properties of curved graphene nanoribbons without extra termination hydrogen atoms are elaborated. The first-principles calculations can simulate their dramatic changes during the unzipping process [59]. The curvature- and width-dependent ground state energies and bond lengths are evaluated, especially for the critical arc angle, critical interaction distance between two-side edge atoms, zipping energy, and unzipping energy. The variation of the band structure and DOS with the curvature is investigated thoroughly. The calculated results clearly indicate the unusual features of energy band, such as energy gap, energy dispersions, band-edge states, mixing bands, band overlap, and state degeneracy. The total and local DOSs exhibit plenty of prominent asymmetric peaks in the inverse of the square-root form. The unzipped and zipped graphene nanoribbons are quite different from each other, mainly owing to the curvature effect and the open/periodical boundary condition. These could be directly verified by the STS measurements.

## 4.1   Curved nanoribbons

A curved graphene ribbon, as shown in Figure 4.1(a), could be regarded as a part of a cylindrical carbon nanotube. Its geometric structure is described by arc angle $\phi$ and curvature radius $R = W/\phi \times 180°/\pi$. $W$ is the nanoribbon width proportional to the number of armchair/zigzag lines along the transverse direction. $\phi$ can account for the nanoribbon curvature, and $\phi = 0°$ and $\phi \neq 0°$ correspond to flat and curved systems, respectively. As for $\phi = 360°$, a $(N_A/2,0)$ zigzag carbon nanotube [a $(N_z/2,N_z/2)$ armchair carbon nanotube; details in the next section] is a rolled-up $N_A$ armchair nanoribbon [a $N_z$ zigzag nanoribbon] in the cylindrical form, where $N_A$ is an even integer. $W$ and $\phi$ will be modulated in the first-principles calculations to simulate the dramatic changes of the orbital hybridizations of carbon atoms on a curved surface.

In the absence of hydrogen passivation, the geometric reconstruction takes place at the ribbon edges. The C-C bond lengths of the edge atoms are much shorter than those of the central ones. Such dangling bonds are strongly dependent on the curvature. For example, a curved armchair ribbon of $N_A = 10$ presents a bond length of ~1.24 Å for the neighboring dimer atoms at the edges within the range of $0° \leq \phi \leq 330°$, as shown in Figure 4.1(b). However, in another range of $330° < \phi \leq 360°$, this bond length grows rapidly with the arc angle until it reaches around 1.41 Å in a zipped surface. The drastic changes in the edge bond lengths clearly indicate the emergences of the significant orbital interactions between the two-side carbon atoms. This range of arc angle, which corresponds to the dramatic transformation of geometric structure, will reveal the critical chemical bondings in determining the transition of electronic properties.

The dependence of the total energy per atom ($E_{tot}$) on the arc angle is useful in understanding the strong competition between the mechanical strain and the interactions of two-side edge atoms. $E_{tot}$ is very sensitive to $\phi$, as shown for various curved armchair ribbons in Figure 4.2(a). $E_{tot}$ grows with the increasing $\phi$, reaches a maximum value at the critical angle ($\phi_c$), and then declines rapidly until $\phi = 2\pi$. All the curved systems exhibit the similar $\phi$-dependence, being categorized into two thoroughly distinct regions: (I) $\phi < \phi_c$ and (II) $\phi > \phi_c$. The total energy is closely related to the bending strain and the covalent bonding. The nanoribbon changes from a flat structure into a cylindrical one during the variation of $\phi$ from $0°$ to $360°$. The enhanced surface curvature leads to the strain energy and thus the increase of the total energy. For $\phi < \phi_c$, the total energy due to the extra chemical bondings is negligible because of the too large distance between two-side edge atoms, in which it is well fitted by [59]

$$E_{tot}(\phi; N_y) \simeq \left[1.25 \times 10^{-6} \times exp\left(\frac{-0.13}{N_A}\right)\right]\phi^2 + \left[\frac{4.9}{N_A} - 10.1\right] \quad (4.1)$$

The first term in the total energy, defined as the strain energy, is inversely

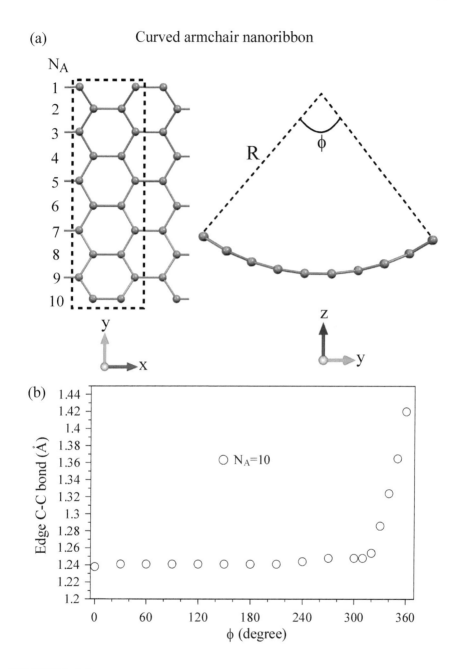

**FIGURE 4.1**
(a) Geometric structure of an $N_A = 10$ curved armchair graphene nanoribbon with radius and arc angle, and (b) the dependence of the edge-C-C bond length on $\phi$.

proportional to the square of curvature (the dashed curve in Figure 4.2(a)). The similar relation is also revealed in various carbon nanotubes [291]. On the other hand, the interaction distance is sufficiently short ($d < 3.4$ Å ) to permit carbon atoms on either side of the armchair ribbon edges to interact with one another, when the arch angle is close to $\phi_c$. The very strong edge-edge atomic interactions, which largely reduce the total energy, compete with the elastic strain to determine the critical angle. The extra orbital hybridizations play a dominating role at $\phi > \phi_c$. In general, zigzag carbon nanotubes have the lower ground state energies, compared with planar armchair nanoribbons with $N_A > 8$. The chemical bondings of carbon atoms in a cylindrical surface are able to suppress the non-negligible mechanical strain.

The physical properties, which are responsible for the zipping and unzipping of a graphene nanoribbon, are worthy of a detailed examination. The critical angle, corresponding to the maximum total energy, grows with the increase of ribbon width (black circles in Figure 4.2(b)). and so does the critical interaction distance (red circles). The stronger chemical bondings between the two-side edge atoms are required to overcome the higher mechanical strain in a narrow curved system, leading to the proportional relationship between $d_c/\phi_c$ and $N_y$. As for the $\phi$-dependent total energy (Figure 4.2(a)), it can define the zipping and unzipping energies in the structural changes. $E_z = E_{tot}(\phi_c) - E(0°)$ accounts for the transformation from a flat nanoribbon into a nanotube. The zipping energy, as shown in Figure 4.2(c) (black squares), presents a monotonous decrease with the increasing $N_y$, directly reflecting the width-dependent strain energy (Figure 4.2(a)). Moreover, the unzipping energy, being required to scissor a closed nanotube longitudinally to become an open nanoribbon, is characterized as $E_{uz} = E_{tot}(\phi_c) - E_{tot}(2\pi)$ (red squares). Except for the very narrow systems, the zipping and unzipping energies have the similar width dependence, in which the former is higher than the latter.

The curved nanographene ribbons, as shown in Figures 4.3(a)–4.3(h), exhibit a lot of features in band structures. The important characteristics cover the asymmetry of valence and conduction bands about $E_F = 0$, diverse energy dispersions, band-edge states, state degeneracy, direct or indirect gaps, and semiconductors or metals, being sensitive to the arc angle. For a flat $N_y = 10$ armchair systems (details in Section 3.1), there are few edge-atom-dominated valence and conduction bands (EAVBs and EVCBs), with weakly dispersive relations (the dotted curves in Figure 4.3(a)). An indirect gap of 1 eV is determined by the $k_x = 1$ state of EACBs and the $k_x = 0$ state of the first valence band ($V_1$ measured from $E_F$). Most of the low-lying energy bands mainly arise from the hybridizations of $2p_z$ orbitals (the $\pi$ bonds at $\phi = 0°$). However, the $2p_y$ orbitals (the $\sigma$ bonds) make significant contributions to the third conduction band ($C_3$), the fourth valence band ($V_4$), and the EAVBs and EVCBs. It is also noted that the $2p_z$ ($2p_y/2p_x$) orbitals on a curved structure can create the $\pi$ and $\sigma$ bonds ($\sigma$ and $\pi$ bonds) simultaneously, owing to the non-parallel and non-perpendicular arrangements [150, 314].

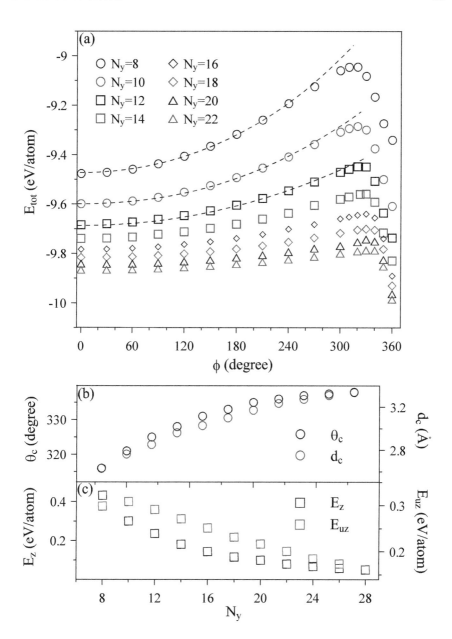

**FIGURE 4.2**

(a) The $\phi$-dependent ground state energy per atom for armchair nanoribbons with distinct widths, (b) the critical arc angles and interaction distances, (c) the zipping and unzipping energies. Also shown in (a) are the strain energies fitted by the dashed curves.

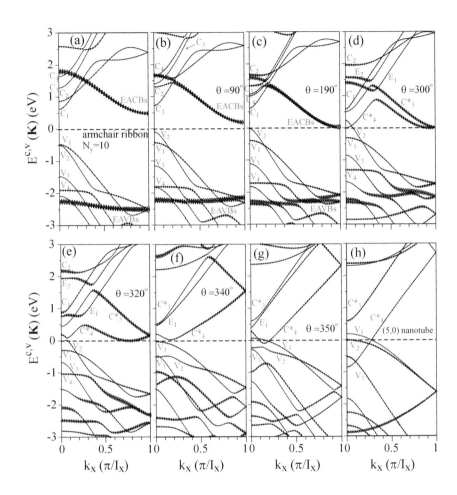

**FIGURE 4.3**
Band structures of the $N_A = 10$ curved armchair nanoribbon under (a) $\phi = 0°$, (b) $90°$, (c) $190°$, (d) $300°$, (e) $320°$, (f) $340°$, (g) $350°$, and (h) $360°$. The radii of open circles represent the contributions due to edge atoms.

Band structures drastically change with the variation of curvature. The asymmetry of energy bands about $E_F$ becomes more obvious in the increase of $\phi$, e.g., $\phi = 90°$ in Figure 4.3(b). The energies, wave vectors, and dispersion relations of the band-edge states are strongly affected by the arc angle. When the arc angle is larger than $60°$, the $k_x = 0$ state of the $V_2$ valence band will become the HOS. Furthermore, the $k_x = 1$ state of the EACBs and the $k_x = 0$ state of the $V_2$ valence band (the LUS and HOS states) approach to the Fermi level simultaneously, leading to a quick decline of an indirect energy gap. Specially, these two bands marginally overlap each other at $\phi \sim 190°$ (Figure 4.3(c)), so that a 1D metal, with a high DOS at $E_F$ due to the band-edge states of parabolic bands, comes to exist. The arc angle in determining the semiconductor-metal transition is relatively small for wider curved nanoribbons. The doubly degenerate states of the EACBs and EAVBs present the significant splittings. Moreover, the band-edge state energies might change the ordering of the conduction/valence bands at $k_x = 0$.

Energy bands are very different from those of the flat nanoribbon (Figure 4.3(a)) for the sufficiently large arc angles, e.g., $\phi = 300°$ in Figure 4.3(d)). The $\sigma$ bondings make more contributions on the low-lying energy bands, compared with the $\pi$ bondings [150, 314]. This directly reflects the significant edge-atom interactions on two distinct boundaries by the $sp^2$ bonding that can create two kinds of free carriers and the significant band mixings. The $V_2$ valence band near $k_x = 0$ crosses $E_F$; therefore, the 1D free electrons and holes coexist in the obvious band overlap. The $C_4$ and $C_3$ bands, which mainly come from the $2p_z$ and $2p_y$ orbitals, respectively, lower their energies and become very close to the Fermi level. The former lies below the latter, since it experiences a stronger curvature effect under the reduced $\pi$-bond contributions and the enhanced $\sigma$-bond ones. The double degeneracy of the EVCBs/EAVBs is largely destroyed, and the split states strongly mix with other energy bands. Specifically, the two EACBs are intermingled with the $C_4$ and $C_3$ conduction bands. The mixed four conduction bands are classified into the $C_4^*$, $C_3^*$, $E_1^*$, and $E_2^*$ bands, based on the predominant orbitals at the $k_x = 0$ state. They have an extra band-edge state in between $k_x = 0$ and 1.

The oscillations of energy bands become very strong, when the arc angle approaches the critical one ($\phi_c = 330°$), e.g., $\phi = 320°$ in Figure 4.3(e). The edge-atom interactions between two different sides are greatly strengthened; therefore, they dominate the quite unstable geometric structure and the unusual band structure. The edge-atom contributions to the $C_4^*$ and $C_3^*$ conduction bands gradually decline. These two bands present higher energies near $k_x = 1$, in which the degenerate states rise from zero to $\sim 0.2$ eV. The $C_4^*$ conduction band moves toward $E_F$, while the $C_3^*$ conduction band exhibits the opposite behavior. The former mixes with the $V_2$ valence band, accompanied with the creation of two extra band-edge states and an energy spacing near $k_x = 0$. Specially, the electronic states corresponding to the previous $C_4^*$ conduction band ($V_2$ valence band) are dramatically changed into those of the present $V_2$ valence band ($V_4^*$ conduction band).

A further increase in the arc angle beyond $\phi_c$ can create the rich and unique band structures, as clearly indicated for $\phi = 340°$ in Figure 4.3(f) and $\phi = 350°$ in Figure 4.3(g). The geometric structure becomes relatively stable, since the $sp^2$ bondings of the two-side edge atoms are initialized. The oscillations of energy bands are greatly weakened, and some band-edge states disappear. The sufficiently large curvatures lead to the hybridizations of the $2p_z$ and $2p_y/2p_x$ orbitals as a result of the shortened distances between the neighboring atoms. These hybridizations cause the $C_4^*$ conduction band to change from an oscillating dispersion into a parabolic one. Mixing of the $C_4^*$ and $V_2$ bands might be absent, and the latter deviates from $E_F$. As to $\phi = 340°$ (Figure 4.3(f)), the LUS and HOS, respectively, correspond to the $C_4^*$ and $V_1$ bands, in which a small indirect $E_g$=0.02 eV represents the semiconducting behavior. A 1D metal is changed into a narrow-gap semiconductor in the range of $\phi_c < \phi < 340°$. This gap will disappear as the arc angle increases to near $360°$, since both C4 and V1 bands overlap with each other (Figure 4.3(g)), i.e., the semiconductor-metal transition reoccurs. The semiconductor-metal transition happens two times during the variation of $\phi$ from $0°$ to $2\pi$. In addition, the $k_x = 1$ degenerate states of the $C_4^*$ and $C_3^*$ conduction bands are considerably raised to $\sim 1.5$–2.0 eV.

A curved armchair ribbon of $\phi = 2\pi$ is a (5,0) zigzag nanotube. The two-side edge atoms bond together along the longitudinal direction to form a very stable hollow cylinder. As a result of the periodical boundary condition, each energy band, as shown in Figure 4.3(h), exhibits a monotonous dispersion relation with a specific transverse quantum number [299, 298]. The band-edge states only appear at $k_x = 0$ and 1. The original edge atoms do not dominate any energy bands (the absence of EACBs and EAVBs), and such atoms make contributions to each one. Most energy bands are doubly degenerate, as predicted by the calculations of the tight-binding model. Some bands are non-degenerate near $k_x = 0$, e.g., the $V_1$, $V_2$, and $C_4^*$ bands. Both $V_1$ and $C_4^*$ bands intersect at the Fermi momentum $k_F \sim 0.25$, in which the former and the latter, respectively, have the free holes and electrons. Furthermore, the $V_2$ band just touches the Fermi level. These can create a high DOS at $E_F$ (discussed in Figure 4.4(e)), clearly illustrating the metallic behavior of a very small (5,0) zigzag nanotube. On the other hand, the tight-binding model predicts it to be a semiconductor in the absence of curvature effects [251]. The large curvatures induce the strong hybridizations of the $(2s.2p_x, 2p_y, 2p_z)$ orbitals (details in Section 4.2); therefore, an obvious energy decrease in the $C_4^*$ conduction band is revealed near $k_x = 0$. The sufficiently strong $sp^3$ bondings are responsible for the metallic carbon nanotubes with very small radii.

The number, intensity, and energy of prominent peaks in DOS strongly depend on the geometric curvature, in which the local DOS can comprehend the contributions due to the edge and central carbon atoms. As to a flat $N_A = 10$ armchair nanoribbon, the low-lying energy bands possess a lot of asymmetric peaks, as shown for (EAVBs, $V_4, V_3, V_2, V_1$) (EACBs, $C_1, C_2, C_3, C_4$) with different symbols in Figure 4.4(a). The strong peaks associated with EAVBs and

EACBs (the black dots) mainly originate from the edge atoms (the dashed blue curves), while the other special structures are dominated by the non-edge ones. Before the critical curvature ($\phi < \phi_c$ in Figures 4.3(b) and 4.3(c)), the number of asymmetric peaks grows with $\phi$, since there exist the non-degenerate EAVBs and EACBs, extra mixed band-edge states and strongly oscillatory energy dispersions. For example, the non-monotonous $C_4^*$ conduction band (the green squares in Figure 4.3(d)) presents four peaks with the reduced intensities near the Fermi level at $\phi = 300°$ (Figure 4.4(c)). In the further increase of $\phi$ (high curvatures in Figures 4.4(d) and 4.4(e)), the peak number declines as a result of the weakened oscillations of energy dispersions and the recovered degeneracy of some bands (Figures 4.3(f) and 4.3(h)). Furthermore, it is very difficult to observe the dominating contributions from the edge atoms; that is, the strong edge-atom bondings will suppress and even destroy the EACBs and EAVBs. A cylindrical carbon nanotube (Figure 4.4(e)), with the transverse periodical boundary condition, exhibits the smallest peak number and the same contributions from all atoms except for two neighboring peaks cross $E = 0$ due to the $V_1$ and $V_2$ bands (the yellow triangle and blue circle). Specially, the peak energies near the Fermi level are responsible for the semi-conducting or metallic behavior. They are related to the $V_2$ band and the EACBs/$C_4^*$ band for $\phi < \phi_c$ (Figures 4.4(b) and 4.4(c); $E_g = 0$), the $V_1$ and $C_4^*$ bands for $\phi_c < \phi < 2\pi$ (Figure 4.4(d); a narrow gap), and the $V_2$, $V_1$ and $C_4^*$ bands for $\phi = 2\pi$ (Figure 4.4(e); zero gap), clearly illustrating the semiconductor-metal transitions during the variation of $\phi$.

The theoretical predictions are worthy of the detailed experimental examinations, especially for the edge-dependent essential properties. Up to now, the curved structures of graphene nanoribbons are confirmed by TEM, SEM, and ATM measurements. The delicate measurements on the curvature, the edge-atom bond length, and the hexagonal arrangement could be further done by STM, as measured for buffer graphene layer [64, 63], graphene bubble [326], and carbon nanotube [364, 256]. On the other hand, the ARPES and STS measurements on electronic properties of curved systems are absent except that the latter have verified the metallic or semiconducting carbon nanotubes. They can examine the curvature-induced properties, covering the conductor-metal transitions, the oscillatory energy bands, the edge-atom-dominated energy bands with weak dispersion relations, and the reduced or recovered state degeneracy. Moreover, such measurements provide the sufficient information in identifying the effects of chemical bondings on the essential properties of curved systems.

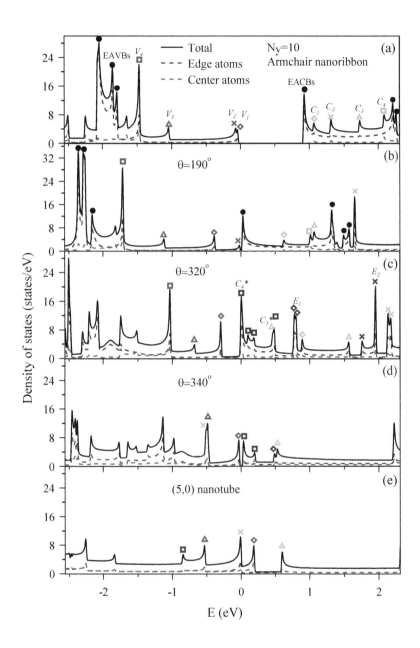

**FIGURE 4.4**
DOSs arising from the total atoms, edge atoms, and center atoms for the $N_A$ = 10 curved armchair nanoribbon under (a) $\phi = 0°$, (b) 190°, (c) 320°, (d) 340°, and (e) 360°.

## 4.2    Carbon nanotubes

Each carbon nanotube, as shown in Figure 4.5(a), is formed by rolling a graphene sheet from the origin to the lattice vector $\mathbf{R_x} = m\mathbf{a_1} + n\mathbf{a_2}$, where $\mathbf{a_1}$ and $\mathbf{a_2}$ are primitive lattice vectors of a 2D sheet. A $(m, n)$ nanotube has a radius $r = b\sqrt{3(m^2 + mn + n^2)}/2\pi$. The chiral angle can characterize the non-chiral/chiral arrangements of carbon hexagons on a cylindrical surface. $\theta = tan^{-1}\sqrt{3}n/(2m + n)$, is the angle between the new $x$-axis (the transverse direction parallel to $\mathbf{R_x}$) and the original $x'$-axis. For all carbon nanotubes, the chiral angles are confined within $|\theta| \leq 30°$. $(m, m)$ and $(m, 0)$, respectively, represent nonchiral armchair and zigzag systems with $\theta = 30°$ and $0°$. It should be noticed that armchair and zigzag carbon nanotubes correspond to zigzag and armchair graphene nanoribbons, respectively. The number of carbon atoms in a primitive unit cell (a rectangle in Figure 4.5(a)) is $N_u = 4\sqrt{(m^2 + mn + n^2)(p^2 + pq + q^2)/3}$, where $(p, q)$ is determined by the primitive lattice vector $(\mathbf{R_y} = p\mathbf{a_1} + q\mathbf{a_2})$ along the nanotube axis. The axial wave vectors in the 1D first Brillouin zone are $|k_y| \leq \pi/|\mathbf{R_y}|$.

There are a lot of theoretical calculations on the electronic properties of carbon nanotubes using the tight-binding model [299, 298] and the first-principles method since the first accurate predictions in 1992 [242, 116]. By the detailed calculations and analyses, all carbon nanotubes could be classified into three types: (I) metals with a finite DOS at the Fermi level, (II) moderate-gap semiconductors with energy gaps inversely proportional to radius, and (III) narrow-gap semiconductors with $E_g \propto 1/r^2$ [150]. When the curvature effects, the misorientation of $2p_z$ orbitals and the $sp^3$ orbital hybridization on a cylindrical surface, are not taken into account, the periodical boundary plays a critical factor in determining the metallic or semiconductoring behavior. The low-energy essential properties are dominated by the $\pi$ bondings of the parallel $2p_z$ orbitals. The electronic states of a 1D carbon nanotube are sampled from those of a 2D monolayer graphene. Carbon nanotubes, with $2m + n = 3I$, are gapless metals, since their linear bands correspond to the Dirac-cone structure, or they can sample the Dirac point from graphene [299, 298]. Furthermore, the $2m + n \neq 3I$ carbon nanotubes are moderate-gap semiconductors. On the other hand, the full orbital hybridizations cause the non-armchair $2m + n = 3I$ carbon nanotubes to become the narrow-gap semiconductors except for $r < 2.5$ Å, and they create the metallic nanotubes with very small radii. In short, the metallic, moderate- and narrow-gap nanotubes are, respectively, characterized by the geometric structures: (I) $m = n$ or $r < 2.5$ Å, (e.g., (5,0) and (6,0) nanotubes) [34, 110]. (II) $2m + n \neq 3I$ ((7,0) and (8,0) nanotubes); (III) $2m + n = 3I$ and $m \neq n$ ((9,0) and (12,0) nanotubes). Three types of energy gaps have been verified by the accurate STS measurements [364, 256, 260]. The $(m, m)$ armchair and $(m, 0)$ zigzag carbon

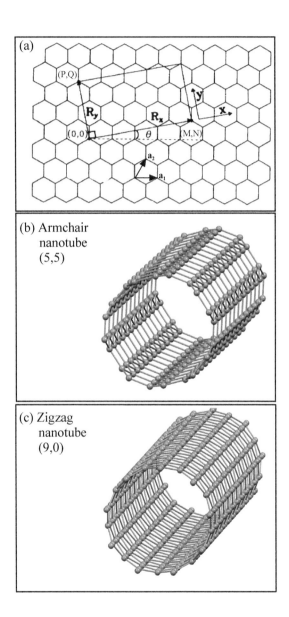

**FIGURE 4.5**
(a) A carbon nanotube, a rolled-up graphene sheet in the cylindrical structure from the origin to the vector $\mathbf{R_x}=m\mathbf{a_1}+n\mathbf{a_2}$, (b) a (5,5) armchair nanotube, and (c) a (9,0) zigzag nanotube.

nanotubes will be chosen for a model study to see three types of electronic structures.

All armchair nanotubes belong to 1D metals even in the presence of significant curvature effects. They have the $k_y$-dependent energy bands with strong dispersion relations, as shown in Figure 4.6 for various armchair systems. The occupied valence bands are asymmetric to the unoccupied conduction bands about $E_F = 0$. Most of energy bands are doubly degenerate because of the cylindrical boundary condition, in agreement with the tight-binding model calculations. This clearly illustrate why the number of energy bands is much lower than that in zigzag graphene nanoribbons (e.g., (3,3) nanotube in Figure 4.6(a)). The band-edge states are situated at $k_y = 0$ and/or near $k_y = 2/3$ (in unit of $\pi/|\mathbf{R_y}|$). The energy dispersions relative to such critical points are in the parabolic or linear form. For larger armchair nanotubes (Figures 4.6(c) and 4.6(d)), there are more low-lying energy band initiated from the $k_y = 2/3$ state. Specially, the first pair of valence and conduction bands, without the double degeneracy, linearly intersect at the Fermi level, indicating the 1D metallic behavior. The Fermi momentum in each armchair nanotube exhibits a red shift relative to $k_y = 2/3$, such as an obvious shift of $\sim 1/6$ revealed by a very small (3,3) nanotube (the dotted-red curved). The former approaches to the latter in the increase of nanotube radius as a result of the reduced curvature effects. Furthermore, for sufficiently large armchair nanotubes, the Fermi velocity, the slope of the linear band, is close to that of monolayer graphene ($3\gamma_0 b/2$; $\gamma_0$ the nearest-neighbor hopping integral of $2p_z$ orbitals) [351].

The effects of curvature on the Fermi-momentum states in armchair carbon nanotubes could also be understood from the tight-binding model. For a planar monolayer graphene, the low-lying energy bands are dominated by the $\pi$ bondings of the parallel $2p_z$ orbitals (Figure 4.6(b)). They present an isotropic Dirac-cone structure near the $K$ point (the corner of the hexagonal first Brillouin zone), as shown in Figure 4.6(b), since the honeycomb lattice has three uniform nearest-neighbor $\pi$ bondings with the same hopping integral ($\gamma_0$). On the other hand, the $2p_z$ orbitals perpendicular to a cylindrical surface are not parallel to one another; furthermore, the angle between two neighboring $2p_z$ orbitals grows with the increasing curvature (or with the decrease of radius), The misorientation of $2p_z$ orbitals can reduce the $\pi$ bondings and create the $\sigma$ bondings (Figure 4.6(b)). By the detailed calculations [207, 282], the hopping integral on the dimer line is different from the other two, depending on the arrangement of three chemical bonds. Even with the modifications in the atomic interactions, the electronic states of armchair carbon nanotubes could sample the $K$ point of monolayer graphene under the periodical boundary condition, being consistent with the first-principles calculations [242, 116]. However, they might induce the obvious changes in the Fermi momentum and velocity. It is also noticed that for very small armchair nanotubes, the $sp^3$ orbital hybridizations make significant contributions to the electronic states near $E_F$ (discussed later in Figure 4.6(d)), while they do not thoroughly alter the metallic behavior due to the linear bands.

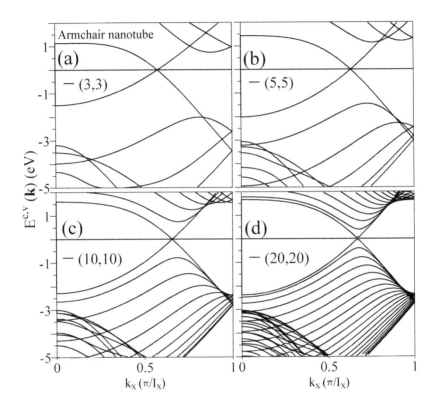

**FIGURE 4.6**
Energy bands of armchair nanotubes: (a) (3,3), (b) (5,5), (c) (10,10), and (d) (20,20).

The electron properties of carbon nanotubes are sensitive to the change in the chiral angle and radius. There are certain important differences between zigzag and armchair carbon nanotubes. All energy bands of $(m, 0)$ zigzag nanotubes, as clearly indicated in Figures 4.7(a)–4.7(d), are initiated from the $k_y = 0$ state with the significant dispersion relations. The low-energy electronic states are doubly degenerate, especially for those in determining the metallic property and the magnitude of $E_g$. Specifically, the $(5, 0)$ and $(6, 0)$ nanotubes, as well as the $(4, 0)$ one [110], present the strong overlap of the valence and conduction bands (Figures 4.7(a) and 4.7(b)); that is, only three very small zigzag nanotubes belong to the 1D metals. This unusual metallic behavior is not dominated by the periodical boundary condition and the $\pi$ bondings, while it mainly comes from the significant $sp^3$ orbital hybridizations on a high-curvature cylindrical surface, as confirmed by the spatial charge distributions and the orbital-projected DOS (Figures 4.9 and 4.10). On the other hand, the $(m > 6, 0)$ zigzag nanotubes possess direct energy gaps at $k_x = 0$. The radius-dependent energy gaps could be further divided into type-II for $m = 3I + 1/3I + 2$ and type-III for $m = 3I$, in which they are, respectively, proportional to the inverse of $r$ and $r^2$, as shown in Figure 4.8 (the dashed red and blue curves).

Specially, the metal-semiconductor transitions appear, when a carbon nanotube is threaded by a uniform axial magnetic field (a magnetic flux $\phi_B$). As a result of the cylindrical symmetry, the $\phi_B$-dependent electronic structure exhibits the well-known periodical Aharonov-Bohm effect, with a oscillation period of flux quantum ($\phi_0 = hc/e$) [5, 229, 212, 211]. For armchair carbon nanotubes, the linearly intersecting valence and conduction bands, with the metallic property, become the separated parabolic ones as $\phi_B$ grows from zero flux. Energy gap linearly reaches a maximum value at $\phi_B = \phi_0/2$ and then declines to zero at $\phi_0$. That is, the metal-semiconductor transitions occur at $\phi_B = I\phi_0$. The similar magneto-electronic properties are revealed in the type-II and type-III carbon nanotubes, in which the critical magnetic fluxes are, respectively, close to $(I \pm 1/3)\phi_0$ and $I\phi_0$. The Aharonov-Bohm effect in carbon nanotubes have been verified from the experimental measurements on the magneto-optical [386, 7], and transport properties [48, 49, 15].

Armchair/zigzag carbon nanotubes are very different from zigzag/armchair graphene nanoribbons in the main features of electronic properties. The edge-atom-dominated energy bands are absent in carbon nanotubes, and each energy band is equally contributed by carbon atoms on a cylindrical surface. The metallic armchair nanotubes (Figure 4.6) sharply contrast with the semiconducting zigzag nanoribbons (Figure 4.7). The latter exhibit the antiferromagnetic spin configuration across the nanoribbon center, and a pair of partially flat bands with the edge-localized states in determining an energy gap. Their band gaps slowly decline with the increment of radius. The $(3I \pm 1, 0)$ and $(3I, 0)$ zigzag nanotubes (Figure 4.8), respectively, have moderate and narrow gaps inversely proportional to $r$ and $r^2$. However, energy gaps in armchair nanoribbons present an inverse relation with width ($W$), in

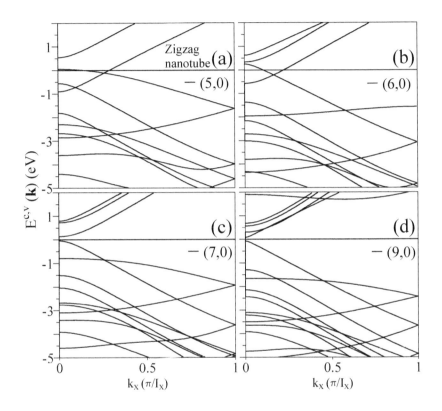

**FIGURE 4.7**
Electronic structures for zigzag nanotubes: (a) (5,0), (b) (6,0), (c) (7,0), and
(d) (9,0).

which those of the $N_z = 3I + 2$ systems are smallest (Figure 3.4). The narrow gaps in zigzag nanotubes and armchair nanoribbons are induced by the non-parallel and non-uniform $\pi$ bondings [150, 323], respectively. Specifically, the metal-semiconductor transitions could be observed in the magneto-electronic properties of carbon nanotubes, but not those of graphene nanoribbons [75]. The above-mentioned differences mainly arise from the boundary condition, the edge structure and the curvature effect. The critical roles, the geometric symmetries and the orbital hybridizations, are responsible for the essential properties.

The geometry- and orbital-dependent DOSs in carbon nanotubes, as shown in Figure 4.10, can provide the useful information about the cooperative/competitive relations among the chiral angle, the periodical boundary condition, and the multi-orbital hybridizations. All armchair nanotubes possess a finite DOS in the plateau form asymmetric about the Fermi level (Figures 4.10(a)–4.10(d)), directly reflecting the metallic behavior of the 1D linear valence and conduction bands (Figure 3.1). However, the metallic zigzag nanotubes presents several prominent asymmetric peaks near $E_F$ ((5,0) in Figure 4.10(e) and (6,0) in Figure 4.10(f)), being associated with the low-lying parabolic energy bands (Figures 4.7(a) and 4.7(b)). For small carbon nanotubes, the low-energy DOSs mainly come from the $2p_z$ and $2p_x$ orbitals (perpendicular to the cylindrical surface and parallel to the tube axis, respectively; Figures 4.10(a), 4.10(b), 4.10(e), 4.10(f) with the red and blue curves); therefore, the strong multi-orbital hybridizations can create the metallic band structures. In addition, the significant hybridization of these two orbitals also has a strong effect on the electronic properties of the highly curved nanoribbons. The contributions due to the $2p_x$ orbitals become weak in the increase of radius; that is, the decrease of curvature leads to the weakened $sp^3$ orbital hybridizations, or the low-energy essential properties are mainly determined by the neighboring $2p_z$ orbitals (the misorientation of $2p_z$ orbitals). The large armchair nanotubes keep the plateau structure across $E_F$ (Figures 4.10(c) and 4.10(d)), while the $(3I, 0)$ zigzag systems present a pair of separated asymmetric peak with a small energy gap (Figures 4.10(g) and 4.10(h)). The chirality and the periodical boundary condition play the critical roles in the metallic/semiconducting property of large carbon nanotubes.

Whether a composite system of carbon nanotube/graphene nanoribbon can exhibit the rich and unique properties deserves a closer discussion. As a result of the rapid advance in chemical analysis techniques, the nanoscale carbon hybrid materials have been successfully synthesized in experimental laboratories, such as, encapsulated $C_{60}$ in single-walled carbon nanotube [320], a 1D carbon chain inside a multi-walled carbon nanotubes [394], graphene/carbon nanotube hybrid [381, 346, 383], and carbon-nanobud hybrid [250]. A commensurate graphene nanotube-nanoribbon hybrid system is formed, when an armchair (zigzag) nanotube is adsorbed on a zigzag (armchair) nanoribbon by the weak van der Waals interactions. According to the first-principles calculations [179, 178], the geometric, electronic and magnetic properties strongly

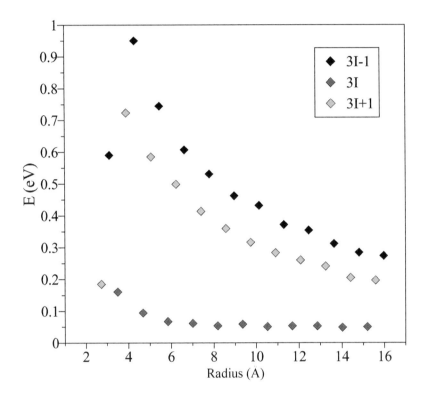

**FIGURE 4.8**
Energy gaps of zigzag carbon nanotubes including $(3I,0)$, $(3I+1,0)$, and $(3I-1,0)$.

**FIGURE 4.9**
The spatial charge distributions for (a) (3,3) and (20,20) armchair nanotubes, and (b) (5,0) and (9,0) zigzag systems.

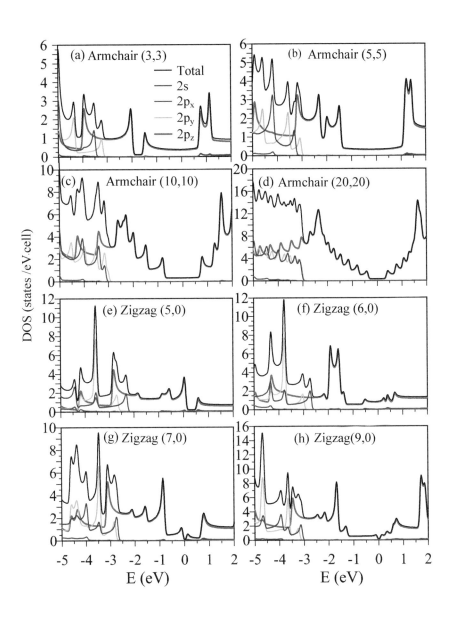

**FIGURE 4.10**
The total and orbital-projected DOSs for (a) (3,3), (b) (5,5), (c) (10,10), (d) (20,20), (e) (5,0), (f) (6,0), (g) (7,0), and (h) (9,0) nanotubes.

depend on the interlayer distance, stacking configuration and spin orientation. The structural stability is examined to be dominated by the charge transfer of $2p_z$ orbitals between two sub-systems. The significant interlayer atomic interactions can greatly modify the low-lying band structures, e.g., the creation of subband spacings and band-edge states, and the destruction of spin/state degeneracy. For armchair nanotube/zigzag nanoribbon hybrids, a pair of intersecting linear bands become two separated parabolic ones with a small energy gap sensitive to the nanotube location. The antiferromagnetic configuration is steadier than the ferromagnetic one for all nanotube locations. Furthermore, the spin degeneracy is lifted when the nanotube is close to the zigzag edge. On the other hand, zigzag nanotube/armchair nanoribbon hybrids exhibit the non-magnetic configuration. The low-energy double degeneracy associated with the cylindrical symmetry is broken by the interlayer $2p_z$-orbital interactions. The predicted essential properties of carbon nanotube-nanoribbon hybrids require further experimental examinations.

The 1D carbon nanotubes, with a unique cylindrical symmetry, are very suitable for studying the basic and applied sciences. The chirality- and radius-dependent semiconductors and metals are available in functionalized nanodevices, such as electronic and optoelectronic devices. The intrinsic 1D band structures, with the decoupled angular momenta, dominate the quantized electrical and optical properties, e.g., the geometry-dependent resistance, capacitance, inductance, and optical conductance [269, 90, 210]. Up to now, the field-effect transistors (FETs) based on the semiconducting carbon nanotubes are widely developed, mainly owing to the advantage of high carrier mobility under the low scatterings. The first FETs composed of single- or multi-walled carbon nanotubes were reported in 1998 [334, 235], which present superior electrical conductance at room temperature using the gate-voltage modulation. A very high carrier mobility ($> 10^5$ cm$^2$/Vs) is identified to be related to the diffusive transport of holes under the electron-phonon scatterings [269, 90]. The nano-scaled radii allows the gate's ability to control the potential of the conducting channel in the ultimate thin FETs, while suppressing short-channel effects [312, 248]. Moreover, the semiconducting carbon nanotubes have the highly potential applications in optoelectronic devices, including electroluminescent light emitters [254, 243], supercapacitors [385], and photodetectors [200, 388, 119]. The band-edge states in 1D nanotubes, with the high DOSs, can create the very prominent optical excitations [210]. The excited electrons and holes on a cylindrical surface are further driven towards each other by using an appropriate bias between source and drain of the FET. The electron-hole recombination will emit the strong electroluminescence. The highly efficient photon-emission process has been extensively utilized in light-emitting diodes (LEDs) [247, 280, 354]. The application of photodetector is based on the electric current generated by the resonant excitations. On the other hand, the metallic carbon nanotubes could be selected as candidates in future interconnect materials of integrated electronic devices [244, 319].

# 5

## Folded graphene nanoribbons

The folded honeycomb lattices may be more complex and intriguing, since they are expected to dramatically modify the essential properties by tuning their structures. In experiments [193, 224, 387, 219, 135], the folded graphene nanoribbons could be produced by the distinct methods, such as, few-layer graphene systems in a chemical solution using the powerful ultrasound [193, 224, 387], the mechanical pressing on a graphene aerogel prefabricated by freeze-drying a homogenous graphene oxide aqueous dispersion and subsequent thermal reduction [219], and the Joule heating of suspended graphene to form the nested structures [135]. The HRTEM measurements clearly show a well-defined zigzag/armchair boundary structure, the coexistent open and closed edges, a nanotube-like curved surface, as well as the AA, AB and other stackings [224, 387]. Furthermore, the semiconducting systems could serve as field-effect transistors with on-off ratios of about $10^7$ at room temperature [193]. On the theoretical side, the tight-binding model and the first-principles method have been utilized to study the essential properties of the folded graphene nanoribbons. The former is available in understanding the unusual quantum Hall effects (the novel magnetic quantization phenomena) [278, 97, 371]. The latter is suitable in evaluating the folding energies, optimal geometries, band structures, energy gaps, density of states, spatial charge distributions, and spin polarizations [177, 382, 396, 60]. It should be noticed that the complex dependences on the edge structures, the nanoribbon widths, the bilayer-like stacking configurations, the van der Waals interactions, the curvatures of folded surfaces, and the spin arrangements need to be included in the calculations simultaneously. This will be explored in detail.

The folded graphene nanoribbons, with the hydrogen passivation, have the unique geometric structures composed of two close open edges, a bilayer-like flat region, and a curved surface (Figure 5.1). The high-symmetry stacking configurations (AA, AA′, and AB stackings) and the various ribbon widths are taken into account to reveal the rich fundamental properties, including the edge-edge distance, interlayer distance, nanotube diameter, folding energy, spatial spin density, band structure, energy gap and DOS. Among the complex geometric structures, we systematically classify four types of armchair and zigzag systems, according to the edge structure, folding energy, number of armchair/zigzag lines, and stacking configuration. Furthermore, the width-dependent energy gaps of the former are further divided into six categories on the basis of the stacking configurations. These eight types of folded systems are predicted to exhibit the diverse essential properties, such as the lower folding energy in the AA′ or AB stackings, the saturated geometric parameters for sufficiently large widths, the metallic linear bands or the separated parabolic bands nearest to $E_F$, the complex width dependences of energy gaps, the creation or destruction of magnetism, and the spin-split energy bands and van Hove singularities. Such features are closely related to the combined effects of quantum confinement, open-edge interaction, van der Waals interaction, curvature, and spin arrangement. The feature-rich folded graphene nanoribbons are expected to provide potential applications in energy materials [27], as well as electronic [193], and spintronic devices [380].

## 5.1    The rich geometric structures

The folded graphene nanoribbons, with the achiral boundaries passivated by the hydrogenated atoms, are chosen for a model investigation. They possess the rich geometric structures. A single flat graphene nanoribbon could be bent into a folded system, in which the latter consists of two flat ribbons connected by a fractional nanotube in the position to generate the curved surface. Figures 5.1(a) and 5.1(b), respectively, correspond to folded armchair and zigzag nanoribbons, accompanied with the hydrogen passivation along the longitudinal $x$-direction (white balls). Each folded nanoribbon, as shown in Figures 5.1(c) and 5.1(d), is characterized by the interlayer distance ($I_z$), nanotube diameter ($D_n$), edge distance ($d_{edge}$), edge structure, stacking configuration, and width. The stable geometric structures are identified from the high-symmetry stacking configurations and the number of armchair/zigzag lines along the transverse direction. There exist three kinds of typical stacking configurations, AA, AB, and AA′ ones (Figure 5.1(e)). Specifically, the AA′ stacking is an intermediate configuration between AA and AB ones, where the C-C dimer of one layer is projected at the hexagonal center of another layer. According to the number of armchair lines, the stable geometric structures could be classified into four types: an even $N_A$ with (I) AA′ (even-aAA′) and (II) AA (even-aAA) stackings (Figures 5.2(a)–5.2(b)); an odd $N_A$ with (III) AA (odd-aAA) and (IV) AA′ (odd-aAA′) ones (Figures 5.2(c)–5.2(d)). The similar configurations for zigzag systems cover an even $N_Z$ with (I) AA (even-zAA) and (II) AB (even-zAB) stackings (Figures 5.2(e)–5.2(f)); an odd $N_Z$ with (III) AA′ (odd-zAA′) and (IV) AB (odd-zAB) ones (Figures 5.2(g) and 5.2(h)). The even-aAA and odd-aAA′ can be, respectively, realized by shifting the upper layers of the even-aAA′ and odd-aAA along the $y$-direction with the distance of $\sqrt{3}\,b/2$. A similar relation exists between the even-zAB and even-zAA stackings (the odd-zAB and odd-zAA′ ones), but with a shift of $b$ ($b/2$). Specifically, for the even-aAA′, odd-zAA′, odd-aAA, and even-zAA stackings (Figures 5.2(a), 5.2(g), 5.2(c), 5.2(e)), the edge of the upper layer lines up

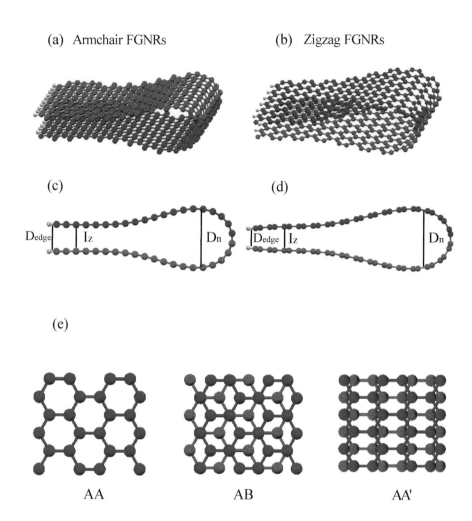

**FIGURE 5.1**

The side view of a folded graphene nanoribbon passivated by hydrogen atoms (white balls): (a) armchair and (b) zigzag systems with (c) and (d) nanotube diameter $(D_n)$, interlayer distance $(I_z)$, and edge distance $(d_{edge})$. (e) shows the top views for the AA, AB, and AA$'$ stackings.

with that of the lower layer, indicating the same $y$ positions for all the edge atoms. Furthermore, the former two only present the $C_2$ rotation symmetry about the $y$-axis, while the others have the $z = 0$-plane mirror symmetry.

The optimal folded structures are achieved under the accurate relaxation simulation. Each stacking configuration exhibits the unique behavior in terms of nanotube diameter, interlayer distance and folding energy. $D_n$ and $I_z$, respectively, reflect the curvature and stacking effects. They, as clearly indicated in Figures 5.3(a) and 5.3(b), are very sensitive to the changes of nanoribbon width and stacking configuration. As for folded armchair nanoribbons, the dependence of $D_n$ on width is weak except that the even-aAA stacking exhibits the decreasing behavior in the increase of $N_A$ and reveals a saturation value at $N_A \geq 40$ (the red circles in the upper part of Figure 5.3(a)). The minimum nanotube diameters of the even-aAA, even-aAA', odd-aAA, and odd-aAA' stackings are, respectively, 7.1 Å, 7.0 Å, 7.0 Å, and 7.2 Å. They are close to that (7.05 Å) of a perfect (9,0) zigzag nanotube. The dependence of interlayer distance on width is negligible. Furthermore, $I_z$ of the AA stackings (the red circles and triangles in the lower part of Figure 5.3(a)) is larger than that of the AA' ones (the green squares and blue trianles), regardless of the even or odd dimer lines. The perceivable $I_z$'s of the former and the latter are 3.5 Å and 3.3 Å, respectively. On the other side, $D_n$'s of folded zigzag nanoribbon initially grow with an increase of $N_Z$ and then reaches a saturation value at $N_z \geq 40$ (Figure 5.3(b)). The maximum nanotube diameters of the even-zAA, even-zAB, odd-zAA', and odd-zAB stackings, respectively, reach 9.2 Å, 9.3 Å, 8.7 Å, and 8.5 Å. They correspond to (7,7) and (6,6) armchair carbon nanotubes, consistent with the experiment measurements [224]. The interlayer distances weakly depend on the ribbon width, in which the saturated $I_z$'s at $N_z \geq 38$, respectively, correspond to 3.5 Å, 3.2 Å, 3.3 Å, and 3.2 Å following the same sequence as for $D_n$'s. The above-mentioned results show that both $D_n$ and $I_z$ are almost constant for the enough large widths. The nanotubes diameters of the folded armchair nanoribbons are smaller than those of the zigzag systems, reflecting the larger deformation (the softer mechanical property) in the former. The similar mechanical-strain response is presented in the deformed carbon nanotubes [186]. The interlayer distances are mainly determined by the stacking effect. The AB and AA' stackings have the smaller $I_z$'s, compared with the AA stacking. This is consistent with those revealed in bilayer graphene nanoribbons [58], and bilayer graphenes [31, 339, 28, 181, 293].

The folding energy, which is required to form a folded graphene nanoribbon from a flat one, is the total energy difference of the former and the latter. $E_{fold}$ consists of the van der Waals energy ($E_{vdW} < 0$), the edge-induced energy ($E_{edge} < 0$) and the bending energy ($E_{bend} > 0$), being contributed by the flat two-layer interaction, edge-edge interaction and mechanical strain, respectively. $E_{fold}$ is greater than zero, since $E_{bend}$ is always larger than $|E_{vdW}| + |E_{bend}|$. $E_{vdW}$ is dependent on the range of the flat region, interlayer distance, and stacking configuration and length of the flat region. This

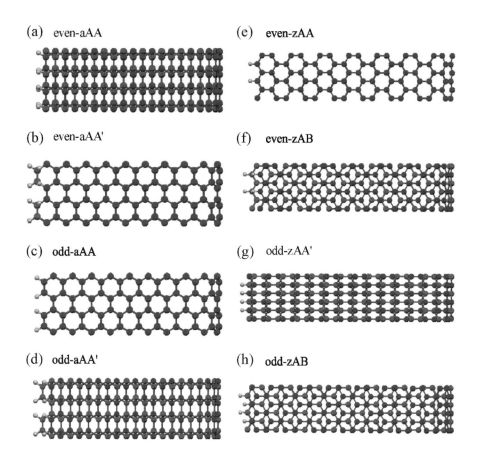

**FIGURE 5.2**
The four types of top views for armchair folded graphene nanoribbons: (a) even-aAA′, (b) even-aAA, (c) odd-aAA, (d) odd-aAA′ stackings, and the similar configurations for zigzag systems: (e) even-zAA, (f) even-zAB, (g) odd-zAA′, (h) odd-zAB stackings.

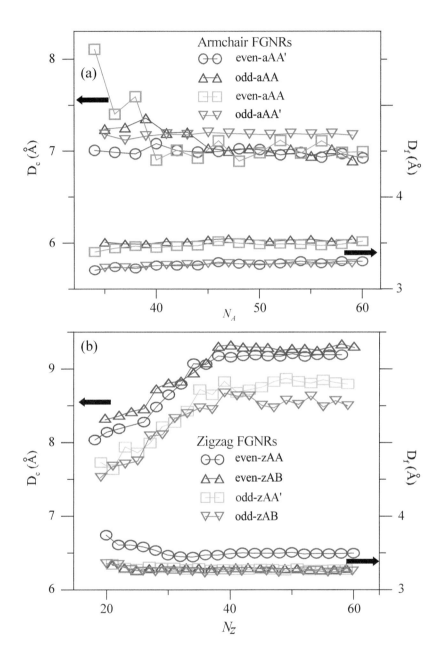

**FIGURE 5.3**
The width-dependent nanotube diameters and interlayer distances of four types of folded (a) armchair and (b) zigzag graphene nanoribbons.

attractive energy is enhanced by the increasing overlap region or the decreasing interlayer distance. Moreover, the magnitude of $E_{edge}$ grows as the distance between two edges declines. As a result, $E_{fold}$ decreases with the increment of width (Figures 5.4(a) and 5.4(b)), mainly owing to the almost constant $E_{edge}$ and $E_{bend}$ (for a specific edge structure and a saturated $D_n$), and the reduced $E_{vdW}$ (for a widened overlap area). The folded armchair nanoribbons exhibit the stacking-dependent behavior, as clearly shown in Figure 5.4(a). $E_{fold}$ is smaller in the AA′ stacking, compared with the AA one. The former is more stable, regardless of the even or odd dimer lines. The edge-edge interactions do not play an important role in the even and odd stackings because of the nearly same inter-edge distance (3.6 Å). On the other side, for the sufficiently wide zigzag systems with $N_z \geq 38$ (Figure 5.4(b)), the folding energies of four types of stackings present the specific descending order: even-zAA > odd-zAA′ > odd-zAB > even-zAB. This indicates the higher stability in the AB stacking configurations. Furthermore, the even-zAB stacking has a lower $E_{fold}$ than the odd-zAB one, since the shorter inter-edge distance in the former (3.1 Å and 3.6 Å in the even and odd systems, respectively) lead to the stronger edge-edge interaction. In addition, the bending energy might account for the difference between zigzag and armchair systems.

## 5.2   The unusual electronic and magnetic properties

The main features of energy bands strongly depend on the quantum confinements, the edge structure, the surface curvature, the stacking configuration and the magnetic configuration. The folded armchair nanoribbons, as shown in Figures 5.5(a)–5.5(h), present a lot of asymmetric valence and conduction bands about $E_F$. The parabolic dispersions are initiated from the $k_x = 0$ state. Some of the band-edge states are doubly degenerate due to the stacking effect. The low-lying bands mainly from the intralayer and the interlayer interactions of $2p_z$ orbitals. The edge-localized energy bands are absent, as observed in planar armchair systems (Figure 3.3). The top of valence band and the bottom of conduction one are located at $k_x = 0$; that is, all the folded armchair systems belong to the direct-gap semiconductors. Apparently, energy gaps change with ribbon width, in which there exists an oscillational relation between $N_A$ and $E_g$ for the AA or AA′ stacking [177]. The similar dependence is revealed in the flat armchair nanoribbons (Figure 3.4). However, the oscillation period of energy gap is, respectively, characterized by six $N_A$'s and three $N_A$'s for the folded (Figure 5.6) and planar armchair systems [323]. For the $N_A=34$ and $N_A=37$ systems, they possess larger energy gaps, and the AA′ stacking is higher (Figures 5.5(a) and 5.5(d)), compared with the AA stacking (Figures 5.5(e) and 5.5(h)).

The folded armchair nanoribbons are different from the planar systems

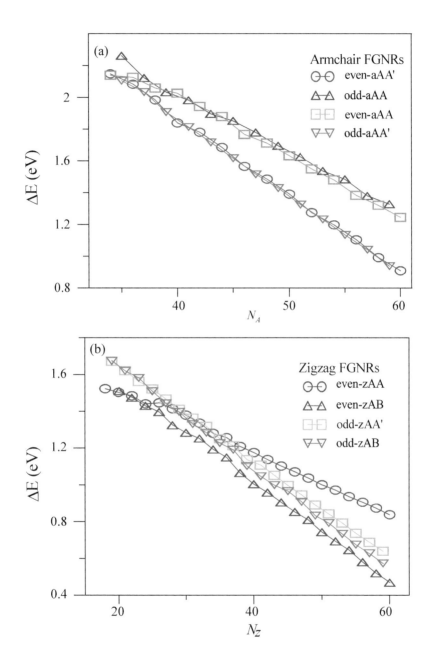

**FIGURE 5.4**
The dependence of folding energy on nanoribbon width for four types of folded (a) armchair and (b) zigzag systems.

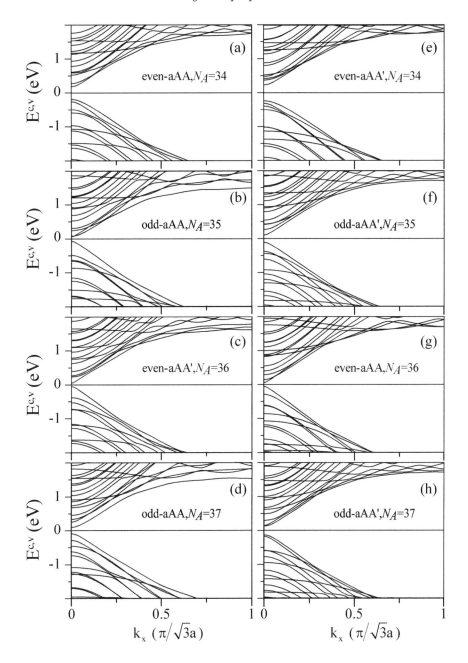

**FIGURE 5.5**
The 1D band structures of the folded armchair graphene nanoribbons for
(a) (even-zAA, $N_A$=34), (b) (odd-zAA, $N_A$=35), (c) (even-zAA,$N_A$=36),
(d) (odd-zAA,$N_A$=37), (e) (even-zAA',$N_A$=34), (f) (odd-zAA',$N_A$=35), (g)
(even-zAA',$N_A$=36), and (h) (odd-zAA',$N_A$=37).

in the width dependences of energy gaps. Based on the AA or AA′ stacking, $E_g$'s of the former could be classified into six categories: $6I$, $6I + 1$,
$6I + 2$, $6I + 3$, $6I + 4$; $6I + 5$. Each category, as shown in Figures 5.6(a) and
5.6(b), exhibits the declining behavior in the increase of ribbon width, reflecting the quantum-confinement effect. The magnitude of energy gap could
be further understood from the even or zigzag dimer lines, as identified from
the three $N_A$-relations. For the even-aAA stacking, it has the largest and
smallest energy gaps for $N_A = 6I + 4$ and $6I$, respectively (red and purple
squares in Figure 5.6(a)). Furthermore, the odd-aAA stacking owns a smaller
gap in $N_A = 6I + 5$ (purple crosses), but the largest energy gap lies in the
$N_A = 6I + 3$ (black crosses) or $6I + 1$ category (green crosses). Similarly,
three categories of width-dependent energy gaps are revealed in the even-
aAA′/odd-aAA′ stacking. The even-aAA′ stacking, with $N_A = 6I + 4$ (red
squares in Figure 5.6(b)), exhibits the largest energy gap among all the armchair systems. The smallest energy gap corresponds to the $6I + 2$ category
(blue squares). The odd-aAA′ stacking for the $6I + 1$ ($6I + 3$) category [green
(blue) crosses] has a larger (smaller) energy gap. The above-mentioned complex width-dependences are attributed to the combined confinement, stacking
and curvature effects in this unique folding structure. Specially, for the flat
armchair nanoribbons, the smallest energy gap is the $3I + 2$ category (Chapter 3), in great contrast with the case for the folded systems. Instead, another
category associated with the smallest energy gap here is $3I$ corresponding to
the even-aAA and odd-aAA′ stackings.

The electronic structure is closely related to the magnetic configuration.
The magnetism of folded graphene nanoribbons is determined by the edge
structure and the edge-edge distance. The armchair systems prefer the non-
magnetic behavior, as observed in planar monolayer ones [323]. Specially, the
spin arrangements in the zigzag systems could be modified by the geometric
structure. The odd-zAA′ and odd-zAB stackings, like planar zigzag nanoribbons, present the anti-ferromagnetic configuration at the open edges, as clearly
illustrated in the spin charge densities (Figures 5.7(a) and 5.7(b)). Each carbon
atom in zigzag edge possesses a magnetic moment of $\sim$0.14 $\mu_B$. On the other
hand, the magnetic configuration is absent in the even-zAA, even-zAB stackings. Whether the magnetism could survive is dependent on the edge-edge
interactions. The strong interactions will effectively suppress the magnetic
moments at a sufficiently short distance. The edge-edge distances of the even-
zAA, even-zAB, odd-zAA′, and odd-zAB stackings are, respectively, 2.9, 3.1,
3.5, and 3.6 Å's, being responsible for the existence of magnetism. The similar magnetic suppression could also be found in bilayer graphene nanoribbons
(discussed later in Chapter 7).

The magnetism and curvature effect will greatly enrich the electronic properties of the folded zigzag nanoribbons, compared with those of the armchair
systems (Figure 5.5) or the planar zigzag ones (Figure 3.9). Most of the energy
bands, as shown in Figures 5.7(c)–5.7(f), present parabolic dispersions. Each
subband might have several band-edge states, in which they are situated at

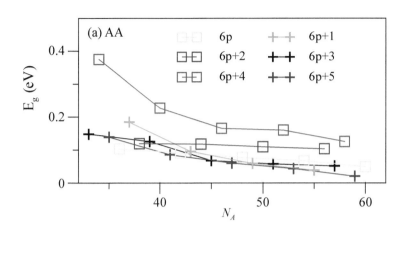

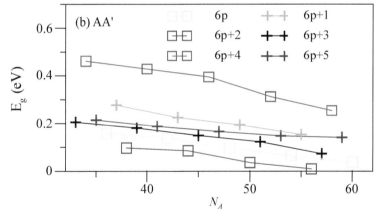

**FIGURE 5.6**

Six categories of width-dependent energy gaps for folded armchair nanoribbons with (a) AA and (b) AA′ stacking configurations.

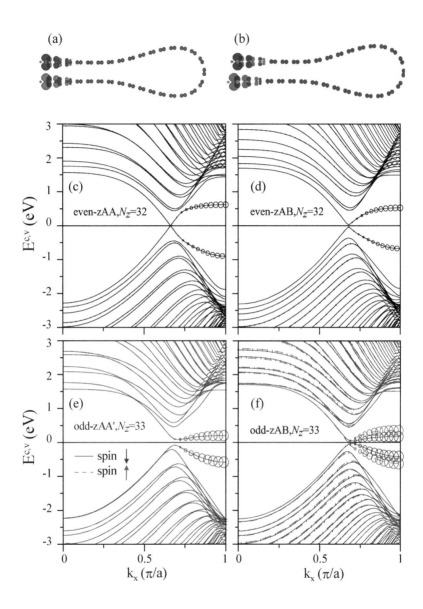

**FIGURE 5.7**
The spatial spin densities of the (a) odd-zAA′ and (b) odd-zAB stackings, and the band structures for (c) (even-zAA, $N_Z$=32), (d) (even-zAB, $N_Z$=32), (e) (odd-zAA′, $N_Z$=33), and (f) (odd-zAB, $N_Z$=33).

$k_x = 0$, 1 and in between them. For the higher/deeper electronic states, there are many anti-crossing subbands with the band-edge states near the zone boundary, mainly owing to the significant interlayer atomic interactions and the enhanced hybridizations of the $(2s, 2p_x, 2p_y, 2p_z)$ orbitals on the curved surface. Specially, the valence and conduction energy bands, which are closest to $E_F$, exhibit the partially flat and parabolic dispersions at $2/3 < k_x \leq 1$ and $0 \leq k_x \leq 2/3$, respectively. The weak energy dispersions mainly come from the local edge atoms (the circle radii representing the edge-atom contributions). The edge-localized energy bands quite differ among four types of folded zigzag nanoribbons in terms of the existence of energy dispersion, band gap and spin splitting.

Both even-zAA and even-zAB stackings exhibit the non-magnetic energy bands, as shown in Figures 5.7(c) and 5.7(d). The former are the only 1D metals, with the linearly intersecting valence and conduction bands at the Fermi level. The metallic behavior is induced by the very strong edge-edge interactions. The Fermi momentum of $k_x \sim 2/3$ is similar to the Dirac point of an armchair carbon nanotube (Figure 4.6), since each folded zigzag nanoribbon has an armchair cylindrical surface characterized by $D_n$. The quasi-Dirac-cone structure is also shown in the AA-stacked bilayer graphenes [125, 367], bilayer zigzag nanoribbons (discussed later in Figures 7.5 and 7.6) and collapsed carbon nanotubes [174, 374], as a result of the mirror symmetry. On the other hand, the latter belong to the narrow-gap semiconductors. A pair of linearly crossing bands become two anti-crossing parabolic bands with a small direct gap. The breaking of mirror symmetry, which comes from the variation of stacking configuration, is responsible for the metal-semiconductor transition, as observed in bilayer zigzag nanoribbons (Figures 7.5 and 7.6) and collapsed carbon nanotubes [174]. As to the antiferromagnetic odd-zAA' and odd-zAB stackings (Figures 5.7(e) and 5.7(f)), they are semiconductors without a gapless or separated Dirac-cone structure. Specially, the double degeneracy of spin degree of freedom is present in odd-zAA' stacking, but absent in the odd-zAB system. Whether the spin-dependent degeneracy is broken depend on the spatial spin arrangements. The $y$-coordinates of the edge atoms on the upper and lower layers are different for the odd-zAB stacking (Figure 5.7(b)), so that the spin-up and spin-down states experience the distinct magnetic environments (red and blue dots). However, the opposite is true for the odd-zAA' stacking.

The width dependences of energy gaps in the folded zigzag nanoribbons are mainly determined by the curvature effect, the quantum confinement, and the magnetic configuration. $E_g$'s of the even-zAB stacking are insensitive to nanoribbon width ($\sim 60$ meV in Figure 5.8 by the purple circles). This reflects the almost constant nanotube diameters (Figure 5.3(b)) and the comparable curvature effect on the separated Dirac-cone structure. In a sharp contrast, band gaps of the odd-zAA' (green circles) and odd-zAB stackings (red and blue troangles) monotonously decline in the increment of width, clearly indicating the finite-size effect. The former are higher than the latter as a result of the absence of spin splitting. For narrow systems ($N_Z \leq 32$), the stacking-dependent order of energy gap is odd-zAA$'$ > odd-zAB with spin down > odd-zAB with spin up > even-zAB. However, the odd-zAB stacking has the smaller gaps for sufficiently wide widths, compared with the even-zAB one. Apparently, the folding structure leads to the diverse $N_z$-dependent energy gaps, compared with the planar zigzag system (Figure 3.9).

The geometric structures and magnetic configurations play important roles in the form, energy, number, and height of van Hove singularities in DOS. All the folded armchair nanoribbons, as shown in Figures 5.9(a)–5.9(d), present a lot asymmetric peaks in the square-root form. Most of them correspond to the $k_x = 0$ state of the parabolic bands (Figures 5.5(a)–5.5(h)). The two adjacent peaks across $E_F$ can determine the width-dependent energy gaps. The main features of DOS is dramatically changed by the edge structure. For folded zigzag nanoribbons, the even-zAA stacking exhibits two prominent peaks on either side of $E_F$ (solid and hollow squares in Figure 5.9(e)), a plateau across $E_F$, and many square-root asymmetric peaks, respectively, arising from the partially flat, linearly crossing and parabolic bands (Figure 5.7(c)). The strong peaks are principally contributed by the local edge atoms. The low-lying constant plateau represents the 1D metal, as observed in armchair carbon nanotubes (Figure 4.6) [298, 242, 116]. However, the even-zAB stacking only has two separate asymmetric peaks near $E_F$ in determining a narrow energy gap (red solid and hollow circles in Figure 5.9(e)). Specially, the anti-ferromagnetic odd-zAA$'$ and odd-zAB stackings possess the spin-up- and spin-down-dependent DOS's (red and blue curves in Figures 5.9(f) and 5.9(g)). The two-component contributions are not distinguishable for the former, but distinct for the latter. That is, the odd-zAB stacking has double low-lying, prominent and square-root peaks. Moreover, regardless of the spin states, the energy spacing of two prominent peaks is smaller in the odd-zAA$'$ and odd-zAB stackings than in the even-zAB one.

The STS/SP-STS measurements could be utilized to identify the types of folded graphene nanoribbons from the various features in DOS's. The zigzag systems possess a pair of strong valence and conduction peaks, a characteristic absent in the armchair ones. The energy and number of the spin-dependent special structures are useful in examining the even or odd zigzag lines. Only the even-zAA stacking has a plateau at $E_F$. The energy spacing of two prominent peaks is smaller for the odd-zAA$'$ and odd-zAB stackings; furthermore, the spin-split double peaks exist in the latter. On the other hand, it is relatively difficult to confirm four types of armchair systems, since the odd-aAA, even-aAA$'$, even-aAA, and odd-aAA$'$ stacking present the similar DOS's. However, energy gaps in DOS are a critical property in identifying the width-dependent six categories. The previous experimental measurements have verified the curvature effect in carbon nanotubes [337], and rippled graphenes [241, 87], and the stacking effect in few-layer graphenes [175]. These two combined effects in folded armchair systems could be further examined by the STS measurements.

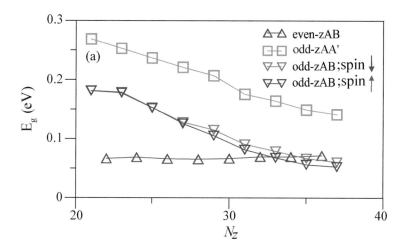

**FIGURE 5.8**
The width-dependent energy gaps of the folded zigzag nanoribbons for the even-zAB, odd-zAA′, and odd-zAB stackings.

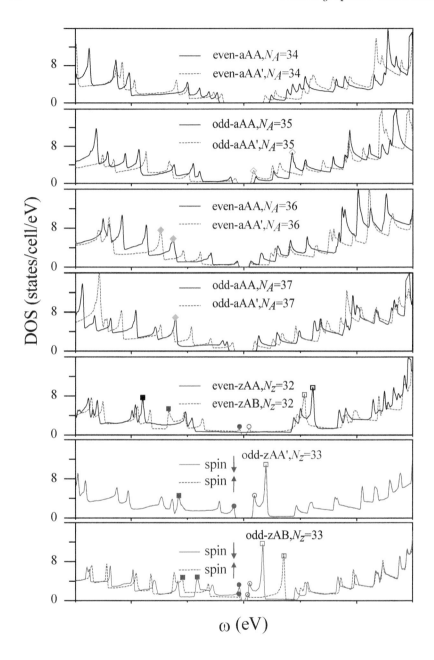

**FIGURE 5.9**
The 1D density of states of the folded graphene nanoribbons for (a) (even-aAA,$N_A$=34) and (even-AA',$N_A$=34), (b) (odd-aAA,$N_A$=35) and (odd-AA',$N_A$=35), (c) (even-aAA,$N_A$=36) & (even-aAA',$N_A$=36), (d) (odd-AA,$N_A$=37) and (odd-AA',$N_A$=37), (e) (even-zAA,$N_Z$=32) and (even-zAB,$N_Z$=32), (f) (odd-zAA',$N_Z$=33) with spin-up and spin-down components, and (g) (odd-zAB,$N_Z$=33) with separate spin contributions.

# 6

## Carbon nanoscrolls

Carbon nanoscrolls and carbon nanotubes could be regarded as the rolled-up graphitic sheets, while they, respectively, possess the open and closed surfaces. Both of them might be produced simultaneously in some experimental syntheses. The scrolled honeycomb structures have been successfully fabricated by the different physical and chemical methods, including the arc discharge [16], high-energy ball milling of graphite [189], chemical route related to intercalation and sonication [345, 305, 311], gas atom absorption on graphene nanoribbons [315], and monolayer graphene predefined with substrates in a chemical solution [370]. The spiral profiles are identified from the HRTEM measurements. The high-quality carbon nanoscrolls could serve as FETs, and they are demonstrated to sustain a high current up to $10^7$ A/cm$^2$ [370]. The open edges and the curved surfaces can provide the available environments in the chemical modifications. Such systems are suitable for both fundamental and applied research. Theoretical calculations were first performed by employing the continuum elasticity theory to analyze the structure and stability of carbon nanoscrolls [338, 176]. However, this approach cannot characterize the atomic-level structural features. The molecular dynamics simulations are available in studying the formation and stability of nanoscrolls [40, 246]. The first-principles calculations predict the high-capacity hydrogen storage after the alkali doping [246], and the strong electromechanical response under the charge injection [295]. The electronic properties show the strong dependence on edge structures and ribbon widths [69, 172]. The roles of quantum confinement, multi-orbital hybridizations, interlayer interactions and magnetic environments need to be clarified in the further calculations. Clearly, carbon nanoscrolls possess the flexible interlayer spaces to intercalate or to be susceptible to doping, indicating the high application potentials in electronic devices, energy storages, Li-batteries, and mechanical devices.

How to form carbon nanoscrolls with the non-uniform curvatures is worthy of a detailed investigation. The first-principles method is suitable in studying the combined effects due to the finite-size confinement, the edge-dependent interactions, the interlayer atomic interactions, the mechanical strains, and the magnetic configurations. The complex mechanisms can induce the unusual essential properties, e.g., the optimal structures, magnetisms, band gaps and energy dispersions. To reach a stable spiral profile, the requirements on the critical nanoribbon width and overlapping length will be thoroughly explored by evaluating the $W$-dependent scrolling energies. A comparison of formation energy between armchair and zigzag nanoscrolls is useful in understanding the experimental characterizations. The spin-up and spin-down distributions near the zigzag edges are examined for their magnetic environments. This accounts for the conservation or destruction of spin degeneracy. The various curved surfaces on a relaxed nanoscroll will create the complicated multi-orbital hybridizations, so that the low-lying energy dispersions and energy gaps are expected to be very sensitive to ribbon width, especially for those of armchair systems. Finally, the planar, curved, folded, scrolled graphene nanoribbons are compared with one another to illustrate the geometry-induced diversity.

## 6.1   The optimal geometries

A carbon nanoscroll is a rolled-up graphene nanoribbon in the open form. It presents a spiral structure in the transverse cross section (the top view of the $y - z$ plane in Figure 6.1), as well as a periodic arrangement along the longitudinal direction (the side view). Its geometry is characterized by two specific edges, an open and curved surface, and a overlapping region with an internal length. The initial and final edges are assumed to be passivated by hydrogen atoms (green balls) in the current calculations. For typical achiral systems, the number of armchair/zigzag lines on the scrolled surface can represent the total and internal lengths (red balls). $(N_A, N_{in})$ and $(N_Z, N_{in})$, respectively, correspond to armchair and zigzag carbon nanoscrolls. The initial ideal structure, as shown in Figures 6.1(a) and 6.2(a), is a perfect arch shape with a specific inter-layer distance. The optimal structures of carbon nanoscrolls are dependent on the initial conditions including ribbon widths and internal lengths (or inner diameters).

After the self-consistent constraint is imposed, an ideal arch structure is gradually changed into an irregular shape. The relaxed scroll geometry, as clearly indicated in Figures 6.1(b)–6.1(d) and Figs. 6.2(b)–6.2(d), is sustained by the layer-layer interactions and simultaneously counterbalanced by the strain forces. The reduced overlapping region caused by the insufficient width will hinder the formation of a carbon nanoscroll. The critical length is closely related to the inner length. Disregarding the periodic edge shape, all the interlayer distances are between 3.22 Å and 3.35 Å, and the average distance is about 3.34 Å. A deeper understanding shows that all the interlayer configurations in carbon nanoscrolls are similar/close to that of the bilayer AB-stacked graphene [31, 339, 28], mainly owing to the higher cohesive energy in the AB stacking. Perceivably, a carbon nanoscroll can be presented as a stable structure, being determined by the sufficiently large width and overlapping length.

The formation of a carbon nanocroll lies in two geometric parameters:

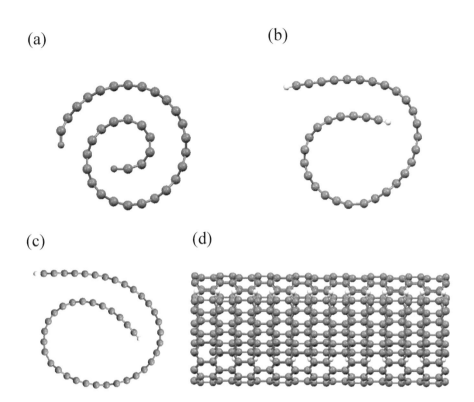

**FIGURE 6.1**
For armchair carbon nanoscrolls, the ideal structure of (a) (34,7) and the optimal structures of (b) (34,7), (c) (43,9); (d) is the side view for the relaxed (34,7) system.

(a)

(b)

(c)

(d)

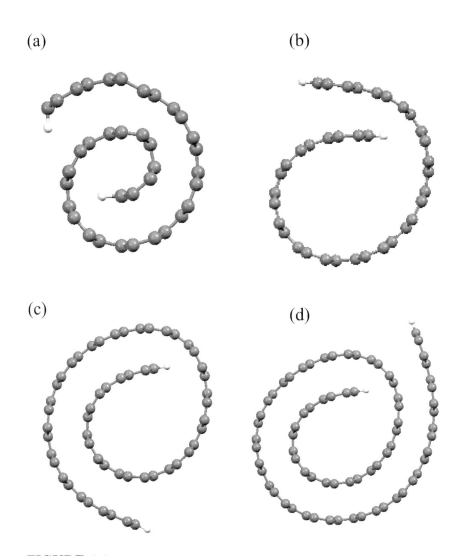

**FIGURE 6.2**
For zigzag carbon nanoscrolls, the ideal structure of (a) (18,4) and the optimal structures of (b) (18,4), (c) (27,4), and (d) (36,4).

the inner length and the scroll width. To hold the structure as a scroll, the required formation energy, the energy difference between the total energy of a carbon nanoscroll and that of a flat graphene nanoribbon, is defined as $E_{scr} = E_{int} + E_{bend}$. $E_{int}$ is the energy arising from the interlayer atomic interactions in the overlapping region; meanwhile, $E_{bend}$ is that due to the restoring force of the mechanic strain. It is obvious that the bending energy is larger than the interlayer interaction energy ($|E_{int}|$), i.e., $E_{scr} > 0$. Roughly, the former is inversely proportional to the square of the effective diameter, as obtained from the various curved nanoribbons (Eq. (1)) and carbon nanotubes [291]. In addition, the significant edge-edge interactions in folded graphene nanoribbons (Section 5.1) are absent in carbon nanoscrolls.

The critical widths and internal lengths rely on the edge structure. Among armchair carbon nanoscrolls, the minimum/threshold width, being examined by the detailed calculations, is associated with the (34,7) system with the critical internal length $N_{in} = 7$. In general, with the increasing nanoscroll width, the overlapping region and the effective diameter grow simultaneously, leading to the enhanced inter-layer atomic interactions and the reduced bending energy. This is responsible for the declining behavior of the scrolling energy, as clearly indicated in Figure 6.3(a). Within the width range of $N_A = 34 - 36$ (blue diamonds), the interlayer distances are relatively large near the end of the overlapping region. The weaker interlayer interactions are reflected in a smaller and smoother variation of the scrolling energy. Concerning $N_A = 37 - 40$, a more stable AB stacking is reached in the optimal structure. Specially, the carbon atoms close to the final edge start to move towards the adjacent layer, so that the scrolling energy decreases more dramatically. In the further increase from $N_A = 41$, the scrolling energy begins with a slow change, but then evolves into a quick decline under the simultaneous cooperation of the interlayer interactions and bending energies. When the critical inner length grows, a wider critical width is required to form a stable nanoscroll. $N_{in} = 9$ and 11 (brown squares and green triangles), respectively, correspond to the minimum widths of $N_A = 43$ and 47. The enhancement of the interlayer interactions cannot effectively compensate that of the mechanical strain. The only way is to reduce the nanocroll curvature by the increment of the threshold width. In considering different inner lengths, it is concluded that all the scrolling energies have a similar dependence on width and a longer inner length corresponds to a lower energy. These results further illustrate the fact that a larger inner diameter results in a smaller bending energy, as discussed earlier.

The zigzag nanoscrolls are similar to the armchair ones in the width-dependent scrolling energy. Specifically, the smallest zigzag system is, as shown in Figure 6.3(b), is (18,4) when the minimum inner length is $N_{in} = 4$ (blue diamonds). The scrolling energy exhibits the wave-like width dependence with the downward trend, mainly owing to the complex competition between the interlayer interactions and the mechanical strains. This indicates that both zigzag and armchair systems share a common dependence during the geomet-

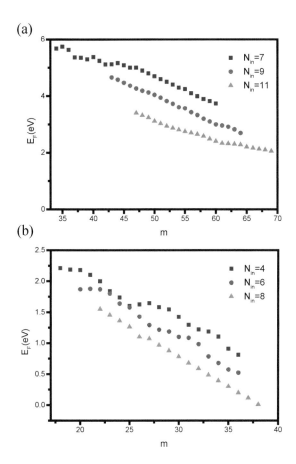

**FIGURE 6.3**
The width-dependent scrolling energies for (a) armchair and (b) zigzag carbon nanoscrolls with the distinct internal lengths.

ric variation. As for $N_{in} = 6$ and 8 (brown squares and green triangles), the threshold widths are $N_Z$ =20 and 22, respectively. Specially, the scrolling energies of zigzag nanoscrolls are much lower than those of armchair systems, mainly owing to the obvious differences in the overlapping region and curvature. This means that the former are relatively easily to be formed and become more stable in the experimental syntheses.

There exist important differences between carbon nanoscrolls and folded graphene nanoribbons in terms of the geometric structures and combined interactions. The former only present the stable AB stacking in the overlapping region, while the latter could reveal the AA, AB and other stackings. The inner diameters grow with the increase of nanoscroll width. However, they reach a saturated value for the wide folded nanoribbons. Such differences are closely related to the combined effects in the unique structures. Carbon nanoscrolls do not have the edge-edge interactions and the distinct magnetic configurations between two zigzag edges. On the other hand, the folding structures possess more complicated interactions. Their width-dependent formation energies exhibit the monotonously declining behaviors (Figure 5.4), when they are characterized by the stacking configuration and the even or odd number of armchair/zigzag lines. That is, the diverse width dependences could be observed in folded graphene nanoribbons.

## 6.2 Electronic properties and magnetic configurations

Band structures are sensitive to edge structure and width, as displayed in Figures 6.4 and 6.5. For armchair nanoscrolls (Figure 6.4), the low-lying energy bands present three distinct types, based on the $N_A = 3I + 2$, $3I + 1$ and $3I$ widths. There are a lot of 1D parabolic bands with the asymmetric valence and conduction energy spectra. They are initiated from the $k_x = 0$ and/or 1 band-edge states. Furthermore, many band-edge states associated with the band mixings are situated at other $k_x$'s, revealing the complex interlayer atomic interactions and curvature effects. Specifically, the $N_A = 3I + 2$ armchair nanoscrolls have a pair of non-monotonous energy bands nearest to the Fermi level, e.g., the (38,7) system in Figure 6.4(a). The highest occupied valence state and the lowest unoccupied conduction state occur at the same wave vector of $k_x = 0.1$, which, thus, leads to a direct energy gap, The similar energy dispersions are revealed in the $N_A = 3I + 1$ armchair nanoscrolls, such as the (40,7) system in Figure 6.4(c). However, such systems have the smaller energy gaps (discussed later in Figure 6.6), compared with the $N_A = 3I + 2$ ones. As for the $N_A = 3I$ armchair systems, they exhibit a pair of weakly dispersive energy bands with the distinct extreme points near $k_x = 0$, e.g., the (39,7) system in Figure 6.4(b). They belong to the indirect-gap semiconductors. It should be noticed that the unusual energy dispersions in determining the direct/indirect gaps are absent in the planar graphene nanoribons (Figures 6.4(d)–6.4(f)).

Electronic structures of zigzag carbon nanoscrolls are enriched by the magnetic configuration. The anti-ferromagnetic spin arrangements is clearly revealed at the initial and final zigzag edges, as shown in Figure 6.5(a) for the (36,4) system. For the distinct spin states, the H-passivated carbon atoms at two open ends experience the different interactions with the surrounding atoms. This is responsible for the four spin-split energy bands nearest to the Fermi level, e.g., the spin-up and spin-down states represented by the red and blue curves in Figure 6.5(b), respectively. Such bands, with the weak energy dispersions, belong to the edge-localized states, as observed in planar zigzag ribbons (Figure 6.5(c)). They will determine two kinds of spin-dependent direct energy gaps. For example, the spin-up and spin-down energy gaps are 0.18 eV and 0.23 eV, respectively. The other energy bands remain the double spin degeneracy, indicating that the distinct magnetic environments only affect the low-lying electronic states. In comparison, the flat systems, with the antiferromagnetic configuration across two zigzag edges (Figure 3.9), present the spin-degenerate energy bands under the equivalent magnetic environment.

The low-lying energy bands are closely related to the orbital bondings on the curved and planar surfaces. Armchair and zigzag carbon nanoscrolls will present the similar and distinct charge distributions (Figures 6.7(a) and 6.7(b)), and so do the orbital hybridizations. The spatial charge density is very sensitive to the surface profile. As for the planar region of the armchair (38,7) nanoscroll, resembling a flat nanoribbon as enclosed by a blue rectangle in Figure 6.7(a), the (2s; 2px; 2py) orbitals of a carbon atom interact with those of the nearby ones to form the $\sigma$ bonds ((I) in orange shades). The perpendicular 2pz orbitals have the significant $\pi$ bondings (dog-bone form by red shades). However, there exist the strong sp3 bondings in the curved regions ((II) by a black rectangle). The similar phenomenon is observed in curved graphene nanoribbon and carbon nanotubes. This leads to the drastic changes in the low-energy band structures, e.g., the unusual small energy gaps and the non-monotonous energy dispersions (Figures 6.4). The four-orbital hybridizations become weaker in zigzag systems, as indicated in Figure 6.7(b), being attributed to the longer C-C distance associated with the projection along the azimuthal direction. The special nanoscroll zigzag structure will induce the different magnetic environments for spin-up and spin-down spatial distributions, and therefore, it creates the spin-spilt energy bands near $E_F$ and the FM configuration.

The main features of DOSs in carbon nanorscrolls are mainly determined the complex cooperation relation among the edge structure, total width, and internal length, as clearly shown in Figures 6.8(a)–6.8(d). The van Hove singularities only come from the parabolic energy dispersions (band structures in Figures 6.4(a)–(c)), leading to the square-root pronounced peaks. The valence and conduction peaks closest to the Fermi level is energy gap corresponding to

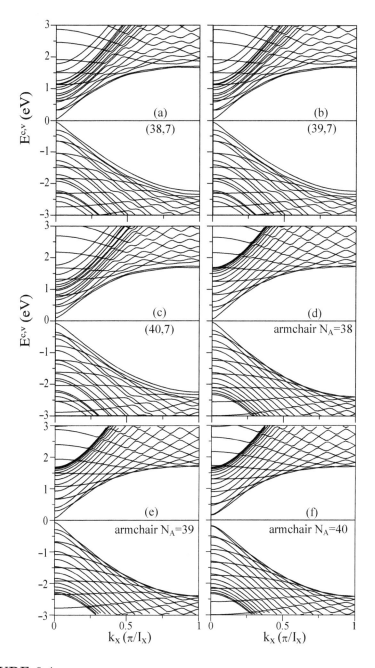

**FIGURE 6.4**

Band structures of (a) (38,7), (b) (39,7), and (c) (40,7) armchair carbon nanoscrolls, and those of (d) $N_A=38$, (e) 39, and (f) 40 armchair nanoribbons.

(a)

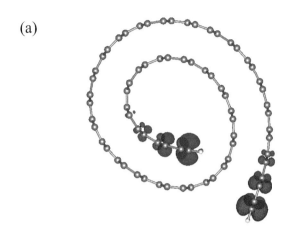

(b)                                              (c)

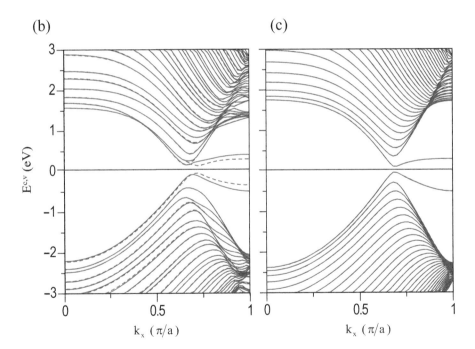

**FIGURE 6.5**

(a) The spatial spin-up and spin-down density distributions in the (36,4) zigzag carbon nanoscroll, indicated by red and blue regions, respectively. Band structures are shown for (b) the (36,4) zigzag carbon nanoscroll and (c) the $N_Z = 36$ zigzag nanoribbon.

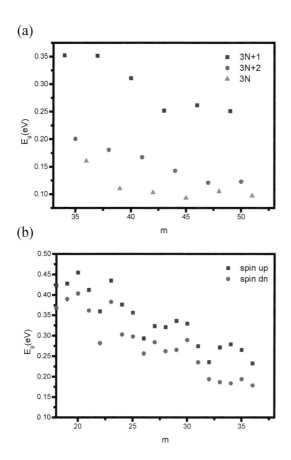

**FIGURE 6.6**
The width-dependent energy gaps for (a) armchair carbon nanoscrolls with $N_{in} = 7$ and (b) zigzag systems with $N_{in} = 4$.

## (a)  Armchair CNS (38,7)

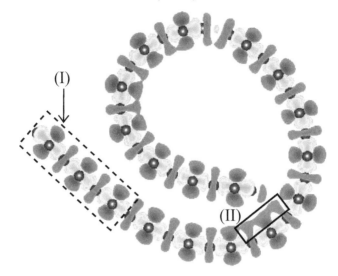

## (b)  Zigzag CNS (36,4)

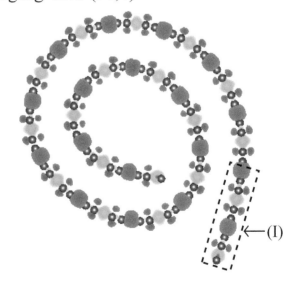

**FIGURE 6.7**
The spatial charge distribution for (a) the (38,7) armchair nanoscroll and (b) the (36,4) zigzag one.

a semiconducting nanoscroll system. The asymmetric peak structures about $E = 0$ are very apparent; furthermore, a simple relation in energy spacing of two neighboring prominent peaks is absent. That is to say, it is very difficult to identify a specific one-to-one correspondence in peak and geometric structures. Both HRTEM and STS need to be utilized to examine the theoretical predictions on the geometric and electronic properties. There are no spin-split peaks in armchair nanorscrolls (Figures 6.8(a)–6.8(c)), while they are present in zigzag systems (blue and red circles in Figure 6.8(d)). The energy splittings, which are due to the partial flat bands at zone boundary (Figure 6.5(b)), are relatively obvious. The SP-STS examinations on them could provide the very useful information on the FM configurations of zigzag nanoscrolls, being in sharp contrast with degenerate behavior from the AFM ones of pristine zigzag graphene nanoribbons.

## 6.3 Comparisons among the planar, curved/zipped, folded, and scrolled systems

The flexible carbon honeycomb lattice can be presented in the various forms under a very strong $\sigma$ bonding. Such structures create the diverse essential properties and thus induce the important differences among the planar, curved, folded, and scrolled graphene nanoribbons. For armchair nanoribbons, only part of the curved systems exhibit the 1D metallic property, mainly owing to the very edge-edge interactions. The similar behavior is revealed in the even-zAA stacking of the folded zigzag systems. The valence and conduction bands, which determines the metallic or semiconducting property, are very sensitive to the geometric structure. All the planar and folded armchair systems have the parabolic bands with direct energy gaps at $k_x = 0$. However, the curved and scrolled ones might possess the non-monotonous energy dispersions with direct or indirect energy gaps (Figure 6.6(b)). Those of zigzag systems belong to the partially flat edge-localized bands at $k_x > 2/3$. An obvious spin splitting appears near the Fermi level when the magnetic environments are different for spin-up and spin-down states near the open edges, e.g., the folded odd-zAB and scrolled zigzag nanoribbons. Specially, only the folded even-zAB stacking presents a pair of linearly intersecting energy band at $k_x \sim 2/3$, as observed in armchair carbon nanotubes.

The width dependences of energy gaps are greatly enriched by the geometric structures. There are three categories in the planar and scrolled armchair nanoribbons (Figure 6.6(a)), but six categories in the folded systems. In addition to $N_A = 3I$, $3I + 1$ and $3I + 2$, the last ones also depend on the odd/even number of dimer lines. For $N_A = 3I + 2$, the planar systems have the smallest energy gaps because of the finite-size confinement. However, the opposite is true for the scrolled systems under the combined effects. In compar-

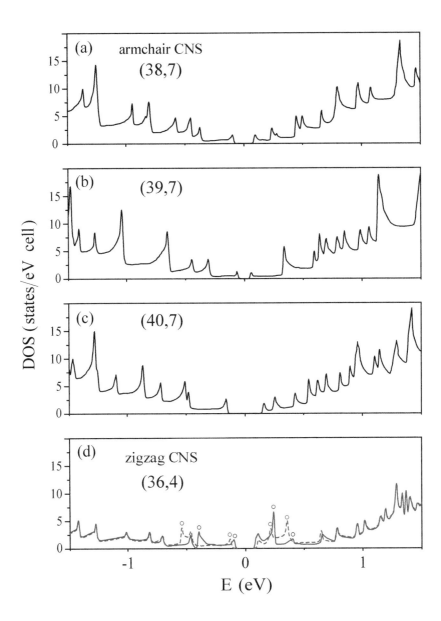

**FIGURE 6.8**
Density of states for the (a) (38,7), (b) (39,7), (c) (40,7) armchair nanoscrolls,
and (d) for the (36,4) zigzag one.

isons among the various pristine systems, the highest energy gaps are revealed in the even-aAA′ folded armchair nanoribbons of $N_A = 6I + 4$. As to zigzag nanoribbons, only the scrolled and odd-zAB folded systems present the spin-split energy gaps. The width-dependent declining behavior is obvious except for the folded even-zAB stacking systems with the strong edge-edge interactions. Furthermore, the wave-like fluctuation comes to exist in the scrolled systems.

# 7

## Bilayer graphene nanoribbons

Few-layer graphene nanoribbons have been synthesized by cutting graphene [17, 193], unzipping multi-walled carbon nanotubes [166, 147], and chemical vapor deposition [361]. They will present quite different essential properties compared with monolayer systems, mainly owing to the number of layers, the various stacking configurations, and the distinct interlayer spin distributions. Specially, the variation of stacking configuration could be initiated by the STM tip to overcome the Van der Waals interactions. Up to now, the electrostatic-manipulation STM is successfully performed on a highly oriented pyrolytic graphite surface [373], graphene/graphite flakes [98, 225], and trilayer graphene [282], clearly indicating the transformation of stacking configuration. The sliding bilayer graphene nanoribbon is expected to be achieved by this method. As for the previous theoretical studies, most of them are focused on bilayer graphene nanoribbons, especially for the typical AA and AB stackings [111, 301, 201, 266]. The calculated results show that stacking configurations play important roles in essential properties. However, the theoretical predictions are not consistent with one another with regard to the existence of magnetism and the planar/non-planar geometry. The controversies might lie in the calculations of the optimal geometric structures. In this work, a thorough study is conducted on bilayer zigzag graphene nanoribbon by varying the stacking and spin arrangements.

Modulating stacking configuration is one of the efficient ways to diversify the essential properties of the layered graphene-related systems. The bilayer zigzag graphene nanoribbons are suitable in understanding the combined effects due to the stacking configuration, edge-edge interaction, finite-size confinement, intralayer and interlayer spin arrangements. The relative shift of the upper and lower layers is predicted to create the dramatic transformations in stacking symmetries and electronic properties [58]. The shift-enriched geometric profiles, edge-edge distances, interlayer distances, energy dispersions, band gaps/band overlaps, band-edge state energies, van Hove singular structures in DOS, magnetic moments, magnetic configurations, and spin-split electronic states are explored in detail. The relatively/most stable configuration, the width-dependent asymptotic behavior, the existence of magnetism, and the metal-semiconductor transition are considered in the delicate calculations. A detailed comparison between bilayer 1D graphene nanoribbon and 2D graphene is also made for illustrating the dimension-dominated diverse properties.

## 7.1  Stacking-enriched geometric structures and magnetic configurations

Bilayer zigzag graphene nanoribbons, with hydrogen passivation (red balls), are chosen to investigate the effects due to the stacking configurations, the van der Waals interactions, and the spin distributions. The various shift-dependent stackings are considered in the calculations. The upper and lower layers, respectively, correspond to green and brown circles, as shown in Figure 7.1(a). Each nanoribbon has two sublattices A and B, indicated by solid and hollow circles, respectively. The four high-symmetry stacking configurations cover AA, $AB_\alpha$, AA′, and $AB_\beta$ (Figures 7.1(b)–7.1(e)). The similar geometric structures could also be observed in folded graphene nanoribbons (Section 5.1). As a result of the 1D finite-width confinement, the AB-stacked systems present two edge-related configurations, $AB_\alpha$ and $AB_\beta$ (Figures 7.1(d) and 7.1(e)). The hydrogen atoms in the AA and $AB_\alpha$ stackings are projected on the same line; however, those in the AA′ and $AB_\beta$ ones are staggered on the distinct lines.

After the detailed self-consistent calculations, bilayer zigzag systems exhibit the rich geometric properties, as clearly shown in Figure 7.2. Both AA and $AB_\alpha$ stackings have the arch-shaped structures in the side view (the $N_Z = 10$ systems in Figures 7.2(a) and 7.2(b)). The upper and lower nanoribbons of the former bend in the opposite directions with a convex profile. However, a wave-like profile, with the same bending manner, is revealed in the latter. The interlayer distances are shorter near two open edges involving twelve carbon atoms, compared with those in the central region. Specially, the degree of bending in terms of the scope is insensitive to the nanoribbon width. The edge-edge atomic interactions are thus expected to play important roles in certain essential properties. On the other side, the AA′ and $AB_\beta$ stackings possess the nearly flat profiles, as observed in monolayer systems. The interlayer distances almost remain uniform from the edge to the central region. Apparently, the arched structures strongly correlate to the interlayer distances. As for (AA, $AB_\alpha$, AA′, $AB_\beta$) stackings, the interlayer distances are, respectively, (3.00, 3.10, 3.72, 3.76 Å) and (3.70, 3.43, 3.71, 3.75 Å) at the boundary and central regions.

The magnetic properties are diversified by the bilayer stacking configurations. The AA and $AB_\alpha$ stackings do not have any magnetic moments on the open zigzag edges, while the opposite is true for the AA′ and $AB_\beta$ ones. Furthermore, the magnetic moments hardly depend on the relative displacement between the former two or the latter two (Path I or II discussed later). This stacking-induced difference could be comprehended from the dependence of magnetic moment on the edge-edge distance, as clearly indicated in Figure 7.3. The former two exhibit a similar $d_{edge}$-dependence (open circles and squares for AA and $AB_\alpha$, respectively). Magnetic moments vanish for $d_{edge} < 3.2$ Å,

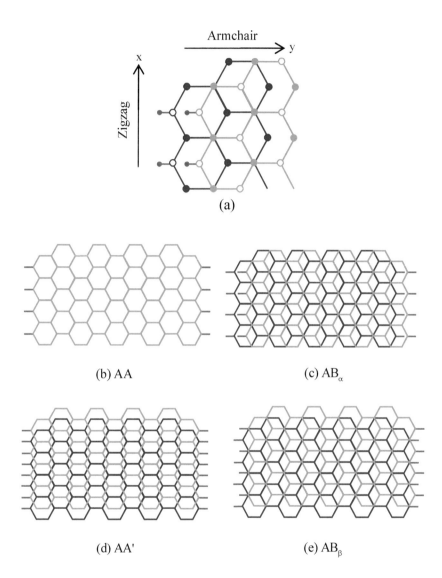

**FIGURE 7.1**
Geometric structures of bilayer zigzag graphene nanoribbons: (a) the relative displacement between the upper and lower layers along the armchair direction, (b) AA, (c) $AB_\alpha$, (d) AA′, and (e) $AB_\beta$ stackings.

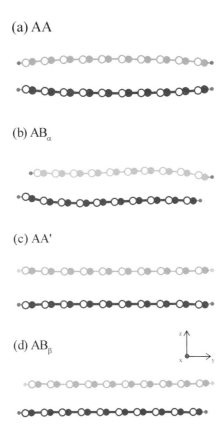

**FIGURE 7.2**
Optimal structures of bilayer zigzag graphene nanoribbons in the side view for (a) AA, (b) $AB_\alpha$, (c) AA', and (d) $AB_\beta$ stackings with width of $N_Z = 10$.

but appear with a strength of 0.15 $\mu_B$ for $d_{edge} > 3.9$ Å. Contrarily, those of the latter two keep a saturated value of ~0.15 $\mu_B$ (solid diamonds and triangles), as a result of the large initial edge-edge distance ($> 3.7$ Å). Moreover, the $d_{edge}$-dependent magnetic moments behave similarly for the four types of spin arrangements in AA′ and AB$_\beta$ stackings. The typical intralayer and interlayer spin arrangements are characterized by (I) AFM-AFM (antiferromagnetic and antiferromagnetic configurations, respectively in Figure 7.3(b)), (II) FM-AFM, (III) AFM-FM, and (IV) FM-FM. Both AFM-AFM and AFM-FM bilayer zigzag nanoribbons, like an AFM monolayer system, are suitable for a model investigation because of the lower ground state energies.

The cohesive energy, the energy difference between the total energy of bilayer graphene nanoribbons and that of two decoupled subsystems, can identify the relatively stable stacking configuration. $E_{coh}$ strongly depend on the relative displacement ($d_y$) between the upper and lower layers, in which two typical paths along the armchair direction (Figure 7.1(a)) are (I) $AA \to AB_\alpha$ and (II) $AA' \to AB_\beta$. Along Path I in the absence of magnetism, the $d_y$-dependent cohesive energy presents the non-monotonous behavior, being classified into into two regions by the critical displacement ($d_{edge} < d_c$ and $d_{edge} > d_c$), as shown in Figure 7.4(a) for the $N_Z = 10$ systems (open circles). $d_c$ (1.06 Å) is smaller than the C-C bond length ($b$=1.42 Å). The bilayer AA stacking, with $d_y = 0$, possesses the largest edge-edge interaction and the smallest stacking effect. With the increasing displacement, the edge-edge interaction and the stacking effect, respectively, declines and grows until $d_y = b$ (AB$_\alpha$). Their strong competition determines the relative stable configuration beyond the AB$_\alpha$ one in the sense that the latter is not the only dominating mechanism. This configuration exists in between the AA and AB$_\alpha$ stackings, but closer to AB$_\alpha$. In another region, the cohesive energy increases with the displacement until $d_y = 3b/2$. The edge-edge interaction is largely reduced and thus the stacking effect makes the major contribution.

Along Path II under the same magnetism, the shift-dependent cohesive energy mainly arises from the composite stacking, edge and spin effects. With the increasing $d_y$, $E_{coh}$ quickly declines until $d_y = b/2$ (AB$_\beta$), and then gradually reaches a minimum at $d_y = 0.84$ Å (Figure 7.4(a)). Furthermore, it starts to rebound and present a local maximum at $d_y = 3b/2$ (AA). The relatively stable configuration is slightly away from the AB$_\beta$ stacking, owing to the smaller energy difference due to the interlayer interactions near the central and boundary regions, and the spin arrangements. Specially, there exists two stable spin configurations, AFM-AFM and AFM-FM, being distinguished by the interlayer spin arrangements. The former has the lower ground state energy at $d_y < b/2$ (open diamonds), in which their energy difference is not significant. The latter becomes more stable in a further increase of shift (open squares). That is to say, a variation of stacking configuration can create a transformation of interlayer spin arrangement.

The combined effects are responsible for the shift process of bilayer zigzag graphene nanoribbon. The relative stable configuration is determined by the

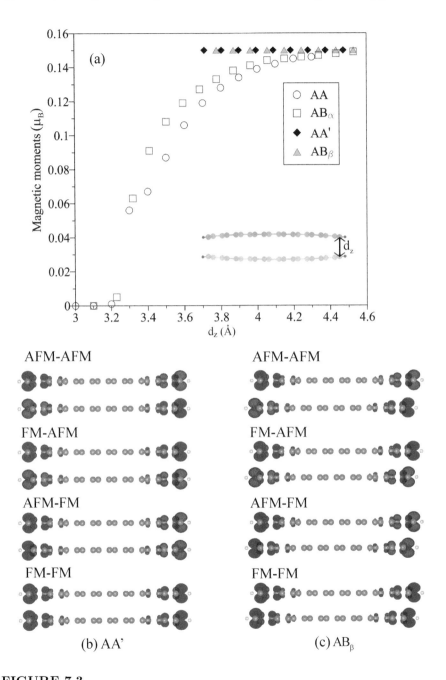

**FIGURE 7.3**
Dependence of the magnetic moment on the inter-edge distance for AA, $AB_\alpha$, AA′, and $AB_\beta$ stackings with width of $N_Z = 10$.

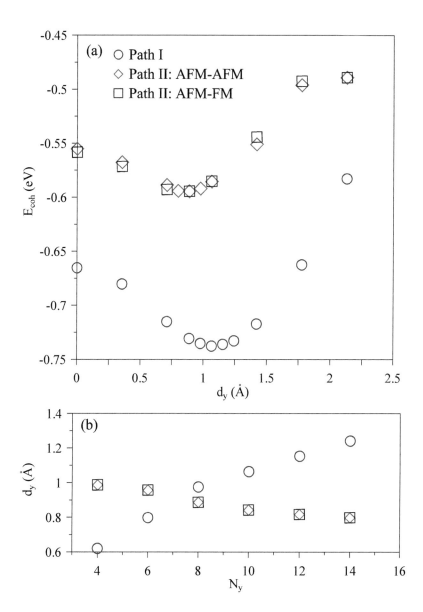

**FIGURE 7.4**

(a) The shift-dependent cohesive energy for the $N_Z = 10$ bilayer zigzag system along AA→AB$_\alpha$ and AA'→AB$_\beta$, and (b) the dependence of the critical shift on the nanoribbon width.

strong competition among the stacking configurations, the edge-edge inter-actions and the spin arrangements. The critical shift ($d_{yc}$), corresponding to this configuration, presents an obvious change when ribbon width grows from $N_Z = 4$, as shown in Figure 7.4(b). Along Paths I and II, $d_{yc}$'s, respectively, approach to $b$ and $b/2$ at sufficiently large widths (open circles and squares/diamonds). This clearly indicates that with an increased width, the configuration of bilayer nanoribbon will change into the AB stacking, as observed in 2D graphene system. The width-dependent asymptotic behavior lies in the enhanced stacking effect, while it hardly depends on the inter-edge interactions and the magnetic configurations. That is, it directly reflects the reduced quantum confinement effect. Specially, for any ribbon widths, the most stable configuration exists between AA and $AB_\alpha$ stackings in Path I, according to the lowest cohesive energy.

## 7.2   Diverse electronic properties

The combined effects can create the rich and unique electronic properties. The band structures near the Fermi level are dominated by the $2p_z$-orbital hybridizations. For the $N_Z = 10$ AA-stacked bilayer nanoribbon, the higher/deeper parabolic bands exhibit the anti-crossing behavior with extra band-edge states as a result of the interlayer atomic interactions. Such interactions also result in two pairs of low-lying conduction and valence bands across $E_F$. Specially, there exist the fourfold-degenerate states (including two spins) near the zone boundary for $E^c = 0.51$ eV and $E^v = -0.85$ eV. These bands are partially flat and mostly contributed by the local edge atoms (open circles), as observed in monolayer nanoribbon. One pair of energy bands linearly intersects at the Fermi momentum, indicating the metallic property with a finite DOS at $E_F$. They behave as those in even-zAA stacking of folded zigzag system (Figure 5.7(c)). Electronic states near $K_F$ are dominated by all carbon atoms, in sharp contrast to the edge-localized states near the zone boundary. The Fermi-momentum states mainly come from the largely enhanced edge-edge interactions under the arc-shaped edge structures (Figure 7.2(a)). This is clearly identified from the loop-like charge density in Figure 7.5(e), especially for the continuous evidences around two boundaries. The significant chemical bonding between two arc-shaped zigzag edges causes the Fermi-momentum states to be similar to the Dirac points in an armchair carbon nanotube (Figure 4.6).

A relative shift between two layers strongly changes the edge-edge interaction, leading to the dramatic transformation of band structure. Two linearly intersecting energy bands become a pair of anti-crossing parabolic bands for various shifts along Path I, as clearly shown in Figures 7.5(b)–7.5(d). Such bands have a direct energy gap between the highest valence state and the low-

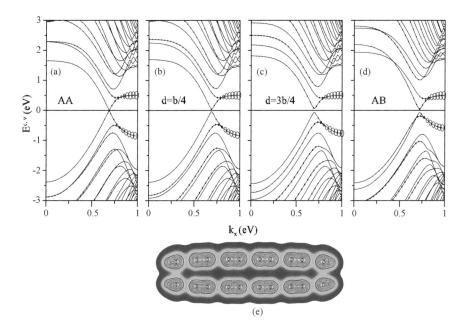

**FIGURE 7.5**
Band structures of $N_Z = 10$ bilayer zigzag nanoribbons under the shifts along
Path I: (a) $d_y = 0$, (b) $b/4$, (c) $b/2$, and (d) $b$. Also shown are the spatial
charge densities for (e) $d_y = 0$.

est conduction state near $k_x = 2/3$. Obviously, the metal-semiconductor transition appears during the modulation of stacking configuration, as observed between the even-zAA and even-zAB stackings in folded zigzag nanoribbons (Figures 5.7(c) and 5.7(d)). This directly reflects the the weakened inter-edge chemical bonding under a larger distance (Figure 7.5(f)). However, the fourfold degeneracy of the low-lying bands near the zone boundary keeps the same. The higher/deeper energy bands are slightly affected by shift, since they are closely related to the intralayer and interlayer atomic interactions.

The distinct magnetic configurations can diversify the electronic structures in terms of band gap and spin degeneracy. The low-lying energy bands in Path II, being composed of parabolic and partially flat dispersions, strongly depend on the interlayer spin arrangements. For the AA′ stacking with AFM-AFM, they are doubly degenerate except for the fourfold-degenerate states near the zone boundary. An indirect-gap semiconductor arises from two anti-crossing bands nearest to $E_F$, as shown in Figure 7.6(a) for $N_Z = 10$ with $E_g^i = 0.2$ eV. Specially, the double and fourfold degeneracies would be broken under the relative displacement, since the environment of the spin-up distribution gradually deviates from that of the spin-down one in the increase of $d_y$ from zero (e.g., Figure 7.3(b)). The AB$_\beta$ stacking (Figure 7.6(b)) exhibits the spin-split energy bands, as observed in odd-zAB folded zigzag nanoribbons (Figure 5.7(f)) and zigzag nanoscrolls (Figure 6.5). The spin-up and spin-down energy gaps are opened up by modulating the stacking configurations. On the other hand, the energy bands of AA′ stacking (Figure 7.6(c)), with AFM-FM, are mostly doubly degenerate and occasionally fourfold degenerate near the zone boundary. The stacking variation does not break the spin degeneracy, reflecting the same magnetic environment of the spin-up and spin-down states under the various shifts along Path II (Figure 7.3(b)). However, the fourfold degeneracy is separated into two doubly degeneracies, and an indirect energy gap might become a direct one.

The band-edge states in bilayer systems appear as the van Hove singularities diversified by the stacking configurations. For the AA stacking, the low-energy DOS, as shown in Figure 7.7(a), presents a plateau structure across $E_F$, two prominent peaks on either side of $E_F$ (solid and hollow squares), and some square-root asymmetric peaks (diamonds and triangles). Such special structures are, respectively, associated with the linear, partially flat bands and parabolic bands (Figure 7.5(a)). A constant DOS at $E_F$ clearly indicates the metallic behavior, as observed in armchair carbon nanotubes (Figures 4.10(a)–(d)) and even-zAA folded zigzag nanoribbons (Figure 5.9). However, the AB$_\alpha$ stacking has a narrow direct gap determined by two separate peaks across $E_F$ (solid and hollow circles in Figure 7.7(b)). The similar structures are revealed in the AA′ stacking with AFM-AFM (Figure 7.7(c)). During the shift from AA′ to AB$_\beta$ stackings, the destruction of spin degeneracy leads to the spin-up and spin-down peaks (red and blue marks in Figure 7.7(d)). The interlayer spin arrangement only has a weak effect on the number and energy of peak structures in the AA′-stacked systems. The AFM-FM and AFM-AFM

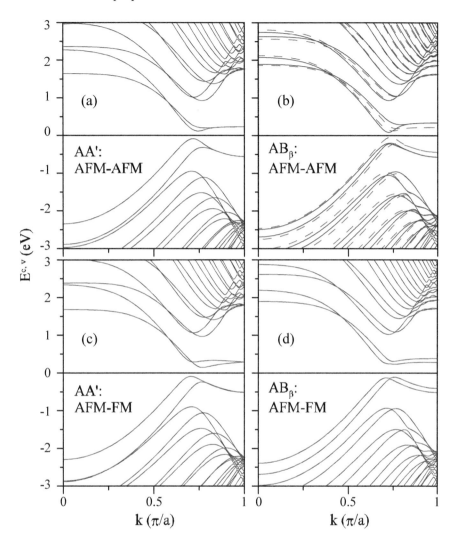

**FIGURE 7.6**
Same plot as Figure 7.5, but shown for (a) AA′ stacking with AFM-AFM, (b) AB$_\beta$ stacking with AFM-AFM, (c) AA′ stacking with AFM-FM, (d) AB$_\beta$ stacking with AFM-FM. The spin-up and spin-down states are indicated by the red dashed and blue solid curves, respectively.

configurations both exhibit the similar DOSs (Figures 7.7(e) and 7.7(c)). With the increase of shift, the number of prominent peaks is drastically changed, e.g., two conduction and two valence peaks in the $AB_\beta$ stacking with AFM-FM (squares in Figure 7.7(f)).

The band-edge states, with the relatively high DOSs, can create the strong perturbation responses and thus dominate the measured physical quantities, such as, tunneling currents in STS, Coulomb excitations, optical absorptions, and electrical conductivities. Their energies ($E_{be}^{c,v}$), as clearly shown in Figures 7.8(a)–7.8(c), rely on the various stacking configurations. They are characterized by circles, diamonds, squares and triangles, based on how close to the Fermi level. The former three arise from the composite bands with parabolic and partially flat dispersions, in which the third ones are due to the edge-localized states near the zone boundary. Furthermore, the conduction and valence states are identified by the hollow and solid symbols, respectively. The energy difference between $E_{be}^c$ and $E_{be}^v$, both being nearest to $E_F$ (hollow and solid circles), is energy gap. As for Path I, $E_g$ increases with $d_y$, then reaches a maximum value at $d_y \sim 1.1$ Å, and finally decreases in the range of 1.1 Å $< d_y < 1.7$ Å (black circles in Figure 7.8(d)). Specially, the spin-up and spin-down state energies are distinguishable in Path II with AFM-AFM (Figure 7.8(b)), e.g., those with the prominent peaks in DOS (red and blue squares in Figure 7.7(d)). The spin-up energy gap (red circles in Figure 7.8(d)) grows with shift until $d_y \sim 0.9$ Å, and declines in the further increase of $d_y$. However, the spin-down one present the opposite dependence (blue circles in Figure 7.8(d)). The spin splitting is absent in Path II with AFM-FM (Figure 7.8(c)), while the degenerate edge-localized state energy might be split under the variation of $d_y$ (black squares). The energy gap exhibits the strongest $d_y$-dependence (green circles in Figure 7.8(d)); that is, $E_g$ possesses the largest value at $d_y \sim b/2$, and is sensitive to a change in the stacking configuration.

## 7.3     Differences between bilayer 1D and 2D systems

The distinct dimensions can greatly diversify the essential properties, being illustrated by the significant differences between bilayer 2D graphene and 1D graphene nanoribbon. When bilayer graphene, as shown in Figure 7.9(a), shifts from the AA stacking along the armchair direction, it gradually approaches to the AB stacking at $d_y = b$, and then it reaches the AA′ until $d_y = 3b/2$. Apparently, layered graphene, with a uniform interlayer distance, presents a planar structure, but not an arc-shaped profile. With the shift of AA→AB→AA′, $I_z$ varies as 3.53 Å→3.35 Å→3.45 Å (solid triangles in Figure 7.9(b)). The highest and the lowest total ground state energies (hollow circles), respectively, correspond to the AA and AB stackings; therefore, the latter is the most stable configuration. They are fully dominated by the interlayer van der Waals

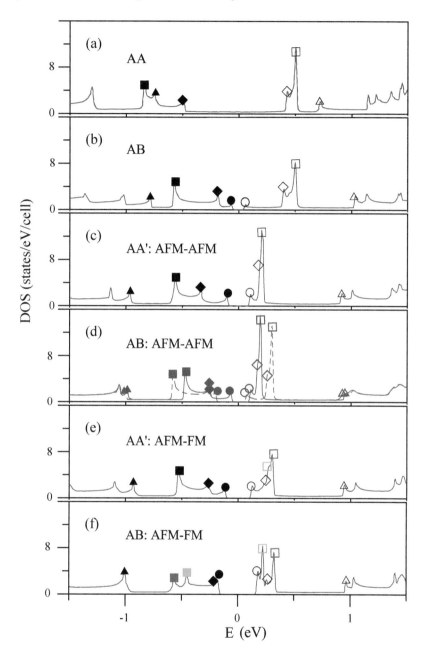

**FIGURE 7.7**
The stacking-dependent DOSs of the $N_Z = 10$ bilayer zigzag nanoribbons for
(a) AA, (b) AB$_\alpha$, (c) AA $'$ with AFM-AFM, (d) AB with AFM-AFM, (e)
AA$'$ with AFM-FM, and (f) AB$_\beta$ with AFM-FM.

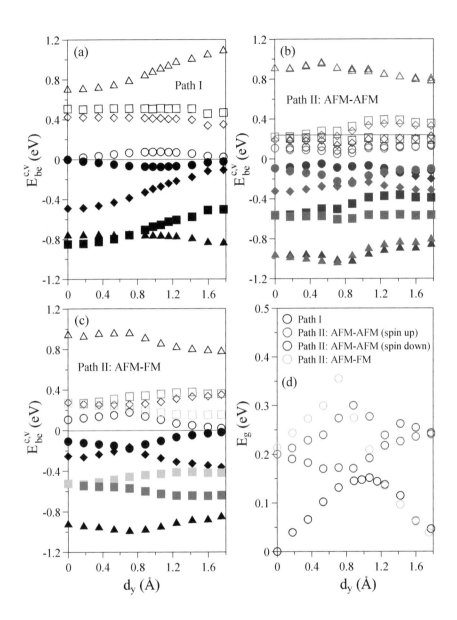

**FIGURE 7.8**

The shift-dependent band-edge state energies of the $N_Z = 10$ bilayer zigzag nanoribbons along (a) Path I, (b) Path II with AFM-AFM, and (c) Path II with AFM-FM. Also shown in (d) are the shift-enriched energy gaps.

interactions, and the edge- and spin-dependent interactions are absent in 2D systems.

Band structures of bilayer graphene are dramatically changed during the variation of stacking configuration. The AA stacking, as clearly displayed in Figure 7.10(a), has two pairs of valence and conduction Dirac cones intersecting at the same K point (the corner of the first Brillouin zone). It is a semimetal with the strong band overlap; that is, there are free holes and electrons, respectively, arising from the valence cone of the first pair and the conduction cone of the second pair. It should be noticed that the free carrier density in bilayer graphene ($\sim 10^{12} - 10^{13}$ e/cm$^2$) is low, compared with that of a 2D metal ($\sim 10^{14}$ e/cm$^2$). The 2D free carrier density, being purely induced by the interlayer atomic interactions of the $2p_z$ orbitals [339, 137], is highest in the AA stacking among all the bilayer systems. With an increased shift, the vertical Dirac-cone structures become distorted, and an arc-shaped stateless region appears near the Fermi level, as shown in Figure 7.10(b) at $d_y = b/8$. This region grows rapidly, and the gapless Dirac points start to separate at $d_y = 4b/8$ (Figure 7.10(c)). After this drastic change, two pairs of well-behaved parabolic bands are formed in the AB stacking ($d_y = b$ in Figure 7.10(d)). The similar effects, the strong mixings of energy bands and the arc-shaped region, are revealed in the further increase of $d_y$ ($d_y = 11b/8$ in Figure 7.10(e)). Furthermore, the Dirac-cone structures are gradually recovered. Finally, two pairs of isotropic cone structures are presented in the AA' stacking ($d_y = 3b/2$ in Figure 7.10(f)), in which they have the tilted axes with the non-vertical Dirac points. They contrast sharply with the vertical Dirac-cone structures in the AA stacking (Figure 7.10(a)).

The special structures of the van Hove singularities in DOS are determined by the dimensions. The low-energy DOS, as clearly indicated in Figures 7.11(a)–7.11(f), is finite at $E_F = 0$ for all the stacking configurations, further illustrating the semi-metallic behavior of bilayer graphene. It is largest in the AA stacking with the strongest band overlap. The bilayer 2D graphene can exhibit the $V$-shape structures, the broadening shoulders, and the symmetric peaks, respectively, corresponding to the isotropic Dirac cones; the extreme points and the saddle ones in the energy-wave-vector space (the band-edge states of parabolic bands) [339, 137, 227]. Specially, both AA and AA' stackings present the coexistent plateau and cusp (Figures 7.11(a) and 7.11(f)), since they are the superposition of two V-shape structures with the different Dirac-point energies (the distinct initiated energies). A composite structure and a pair of shoulders are revealed in the 11$b$/8 stacking (Figure 7.11(e)), reflecting the complicated low-lying bands. The latter also appears in the AB stacking (Figure 7.11(d)). Specially, the symmetric peaks, being associated with the highly distorted energy bands, could be observed in the $b/8$ and $4b/8$ stackings (Figures 7.11(b) and 7.11(c)).

There are a lot of significant differences between bilayer graphene and graphene nanoribbon. The finite-size confinement, edge structure, and spin distribution are absent in the former, and so do the non-planar structure,

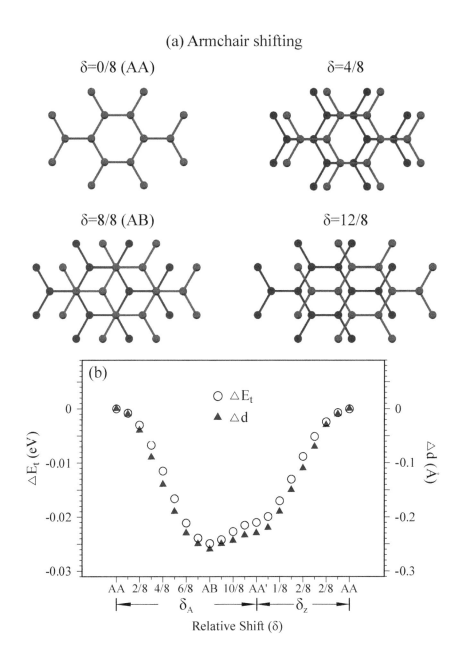

**FIGURE 7.9**
(a) Geometric structures of bilayer graphene shifting along the armchair direction and (b) their ground state energies and interlayer distances.

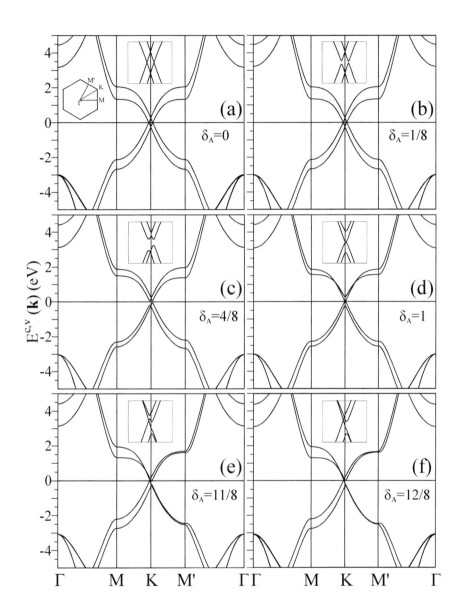

**FIGURE 7.10**
The shift-dependent band structures for bilayer graphenes with (a) $d_y = 0$, (b) $b/8$, (c) $4b/8$, (d) $b$, (e) $11b/8$, and (f) $3b/2$.

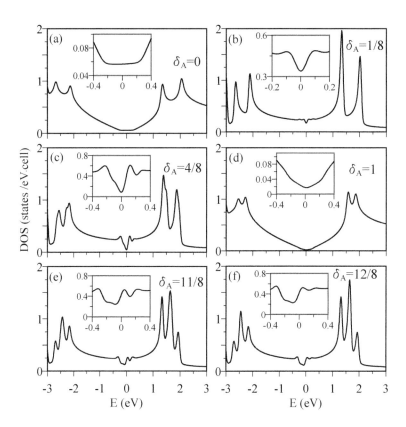

**FIGURE 7.11**
The shift-dependent DOSs for bilayer graphenes with (a) $d_y = 0$, (b) $b/8$, (c) $4b/8$, (d) $8b/8$, (e) $11b/8$, and (f) $12b/8$.

edge-dependent chemical bonding, partially flat edge-localized bands, spin-split electronic states, energy gap, and magnetic moment. The most stable configuration of the former is the AB stacking, while that of the latter deviates from it. The former and the latter belong to semimetals and semiconductors/metals, respectively. Most of 1D nanoribbons present the direct or indirect energy gaps. Specifically, the linear bands in 2D and 1D systems, respectively, create the semi-metallic and metallic behaviors as a result of dimension-dominated carrier density. Graphene nanoribbons might have the intralayer and interlayer spin arrangements near the zigzag edges, leading to the magnetism-enriched band structures and DOSs. The special structures in 2D and 1D DOSs, respectively, appear as a composite plateau and cusp/ shoulders/symmetric peaks and a plateau/asymmetric peaks/prominent peaks, being induced by the critical points in the energy-wave-vector spaces.

# 8

# *Edge-decorated graphene nanoribbons*

The open edges of graphene nanoribbons, with the dangling bonds, are available in modulating the essential properties by the adatom decorations. The strong chemical bondings between adatoms and carbons might reconstruct the edge structures and thus dramatically change the electronic properties. During the chemical synthesis of graphene nanoribbons, the monoatomic molecules could be attached to the edge carbon atoms, e.g., H [330], and O [166]. Furthermore, the edge chlorination of graphene systems has been achieved under the specific chemical methods [333]. The H-decorated $N_A = 7$ armchair nanoribbon, with planar and open edges, is identified from the STM measurements and simulations. The oxygen passivation appears in the unzipping of carbon nanotube using $KMnO_4/H_2SO_4$, in which the O-C bonds are examined by the XPS measurements. The open and distorted edge structures, with the stacking configurations, in the chlorinated graphene nanoribbons and nanographenes are confirmed by the X-ray single-crystal diffractions. Furthermore, the XPS analyses present the strong evidences of the Cl-C bonds. In addition to the well-known H adatoms, some elements are predicted to be stable in the edge passivation, such as K [234], F [133, 228], O [131, 268], B [105], Mg [105], Ru [302], Te [14], and transition metal [356]. Such studies are mainly focused on the modulation of electronic properties under specific planar configurations. These edge-decorated systems may be 1D metals or semiconductors. There also exist a few investigations done for the most stable configurations of the edge-decorated systems [317, 348, 217]. When the curved and flat graphene nanoribbons are successfully produced through chemical methods by unzipping carbon nanotubes [166, 51, 47, 167, 310], the adatom decorations might occur simultaneously during the strong reaction processes. The curved structures and dangling bonds provide a very suitable environment for modulating the geometric structures and electronic properties. The previous calculations show that the combined effects arising from the curvature and adatom passivation play critical roles in the diversified essential properties [57], especially in the tubular/curved/flat structures and the metallic/semiconducting behaviors. Apparently, a systematic study is indispensable in understanding the chemical functionalization of nanomaterials.

How to diversify geometric structures and electronic properties by the edge passivation could be thoroughly investigated using the first-principles calculations (Section 8.1). The decoration adatoms, with atomic numbers lower than twenty in the periodic table, are chosen to make a systematic study, since they tend to form the most stable configurations. Many symmetric configurations of adatom-decorated armchair nanoribbons are examined to achieve the most stable one. The X-C and X-X bond lengths, planar or nonplanar structures, adsorption energies, charge transfers, energy gaps, band structures, and density of states are evaluated in detail. How many types of geometric structures are obtained from the various adatom decorations. The edge-carbon- and adatom-dominated energy bands, the adatom- and C-orbital-decomposed DOSs, and the edge charge distributions are useful in understanding the decoration effects due to the significant C-C, X-C and X-X bondings, e.g., the drastic changes in the width dependences of energy gaps and the special structures of DOS.

The adatom decoration during the unzipping of carbon nanotube can present the edge-passivation and curvature effects and thus the rich essential properties (Section 8.2). Armchair carbon nanotubes and highly curved/planar zigzag graphene nanoribbons are suitable for a model study. Decorating adatoms, which have atomic number of lower than 30 and are capable of bonding with edge carbon atoms, are used to systematically investigate the optimal geometric structures, deppending on the X-C distances, X-X distances, charge transfers, and edge-atom interactions. The magnetic configurations in zigzag edges and 10 transition metal elements are taken into consideration. The bending strains induced by the curvature increase the total energy, while the edge-edge interactions reduce or enhance the total energy. The competitive or cooperative relationship between these two effects that decides the resulting decorated structure will be explored in detail by the simulation model. The curvature effect and the decorating atoms will give rise to the feature-rich electronic properties, especial for the low-lying energy band and the metallic/semiconducting behaviors. The DOSs at the Fermi level are compared with those of metallic carbon nanotubes, and they might be potentially important in various applications.

## 8.1   Adatom edge passivations

The $N_A = 8$ armchair graphene nanoribbon is chosen as a representative system, in which an unit cell contains 4 adatoms and 16 carbon atoms. The armchair boundaries are relatively easy in creating the edge reconstruction, compared with the zigzag ones. Elements, with an atomic number of less than 20, can serve as the edge-passivated adatoms to fully illustrate the decoration effects. An obvious charge transfer, being induced by the significant chemical bondings between the edge carbon atoms and the adatoms, is evaluated from the Bader analysis method. Moreover, the adsorption energy, the reduced total energy due to the  adatoms bondibg with the edge carbons, is characterized as the total energy difference between an edge-decorated graphene nanoribbon and two independent sub-systems (pristine nanoribbon and isolated adatoms). Specially, the former might have part of the total energy arising from a mechanical strain. In addition, the delicate calculations on the various total energies are done for the same unit cell.

With the detailed relaxation calculations, a lot of symmetric configurations have been examined for distinct passivation atoms. The most stable structure, as clearly illustrated in Figures 8.1(a)–8.1(c), corresponds to each adsorbed element in the state of the lowest total energy. According to the planar distortions/reconstructions and the edge-atom arrangements, the optimal geometric structures could be classified into three types: type I, type II, and type III. The main features, the geometric profiles, charge distributions and charge transfers, are discussed as follows.

Adatoms, which can create the planar edge-decorated systems, cover H, N, Be, and B (Figure 8.1(a)). The H-decorated nanoribbon has a regular honeycomb lattice with the H-C bond length of 1.08 Å. A tiny electric charge of ~0.02 e is transferred from H towards the outermost C atom (Table 8.1), revealing a covalent H-C bond. Furthermore, this bond only appears as a protruding edge structure, but not one part of a polygon. That is to say, the H adatoms cannot bridge the strong interactions of carbon atoms on the neighboring edge dimers. Apparently, the 1s orbital is not sufficient in creating the multi-chemical bondings. On the other hand, the unusual bridging effects are clearly revealed in in the other type-I systems as the distinct closed profiles (the different polygons).

The N-decorated graphene nanoribbon presents a periodic heptagon-pentagon edge structure, accompanied with the regular hexagons (Figure 8.1(a)). A significant reconstruction occurs after edge absorption, in which this composite profile has the non-uniform chemical bonds. The N-C bond lengths in the heptagons and pentagons are, respectively, 1.30 Å and 1.28 Å (Table 8.1). However, the C-C bond length near the ribbon center agrees well with that of pristine graphene ($\sim 1.42$ Å). N has the higher electronegativity, compared with C. More electrons are distributed closer to the adatoms, as clearly illustrated by a strong covalent bonding in Figure 8.2(a) (red region). Each N adatom is estimated to gain extra 3.13 e and 2.77 e in heptagonal and pentagonal regions, respectively. Specially, the Be adatoms can bond with the C atoms at the same and neighboring dimers simultaneously, so that a closed triangle-pentagon edge structure is formed at each edge. The original honeycomb lattice almost keeps the same. The Be-C bond lengths are 1.73 Å and 1.85 Å in the triangular and pentagonal parts, respectively. Each Be adatom contributes two outmost valence electrons to the bonding carbon atoms (2 e in Table 8.1). The full charge transfer indicates the highly ionic behavior of adatom. As for the B-decorated system, the non-uniform pentagons, with B-C and C-C bonds, appear at the edges. The B-B bonds, with a length of 2.00 Å, play a critical role in creating the closed edge structures. Each B adatom transfers $\sim 1.48$ e to the neighboring C atom, indicating a significant covalent bonding. As for the type-I systems, the unusual adatom-dependent

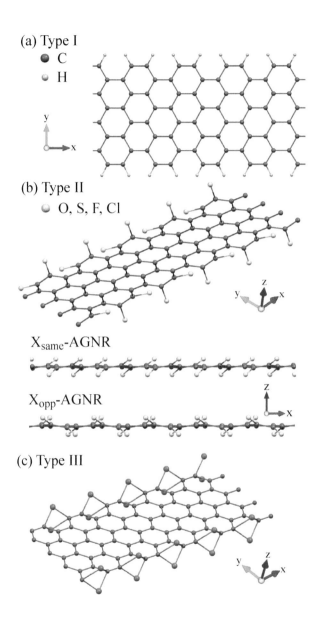

**FIGURE 8.1**
Geometric structures of (a) type-I, (b) type-II, and (c) type-III edge-decorated graphene nanoribbons.

## TABLE 8.1

The edge passivation of $N_A = 8$ armchair graphene nanoribbon using adatoms with an atomic number less than 20, in which the boundary reconstructions present the various characteristics: type-I, type-II, or type-III edge structures, X-C bond length, X-X bond lengths, absorption energies, charge transfers, and semiconducting or metallic behaviors.

| Type | Adatom | X-C bond (Å) | X-X bond (Å) | $\triangle E$ (eV) | Charge transfer (e)* | (Semi-) conductor |
|------|--------|--------------|--------------|--------------------|----------------------|-------------------|
|      | H      | 1.08         | 1.83         | -4.359             | 0.02                 | S                 |
| I    | N      | 1.30 / 1.28  | 2.12         | -5.881             | 3.13                 | S                 |
|      | Be     | 1.85 / 1.73  | 2.15         | -3.707             | 2.00                 | M                 |
|      | B      | 1.57         | 2.23         | -5.500             | 1.48                 | S                 |
|      | O      | 1.22         | 3.03         | -5.776             | -1.73                | S                 |
| II   | S      | 1.75         | 2.03         | -4.274             | 0.09                 | S                 |
|      | F      | 1.34         | 2.58         | -4.795             | -0.78                | S                 |
|      | Cl     | 1.71         | 3.21         | -3.276             | -0.16                | S                 |
|      | Li     | 2.01         | 2.38         | -2.216             | 0.83                 | M                 |
|      | Na     | 2.49         | 3.06         | -0.732             | 0.59                 | M                 |
|      | K      | 2.81         | 3.86         | -1.413             | 0.51                 | M                 |
| III  | Ca     | 2.39         | 3.38         | -0.819             | 0.90                 | M                 |
|      | Mg     | 2.25         | 2.78         | -2.011             | 1.05                 | M                 |
|      | Al     | 2.14         | 2.78         | -3.558             | 1.60                 | S                 |
|      | Si     | 2.15         | 2.23         | -3.483             | 1.16                 | S                 |
|      | P      | 1.83         | 2.23         | -3.732             | 0.50                 | S                 |

*The plus or minus signs, respectively, denote that electrons are gained or lost.

edge morphologies are closely related to the atomic radii and the number of valence electrons.

Type-II systems present the open and distorted edge structures, as predicted in an oxygen-decorated system [317, 217]. Their total ground state energies are mainly determined from the competition of the edge bonding and the mechanical distortion. According to the lowered total energies (Table 8.1), the former can effectively suppress the latter. There exist two stable configurations, namely, $X_{same}$- and $X_{opp}$-decorated armchair nanoribbons (Figure 8.1(b)), compared to the higher-energy planar configuration. X represents a certain kind of adatom. The X-C bonds exhibit the opposite distortions in the periodic form along the edges. As to the corresponding X-C bonds placed across the nanoribbon center, they are located at the same $z$-position for the former, while the opposite is true for the latter. Specially, these two configurations only have a small total energy difference of a few meVs. The lower-energy $X_{same}$-decorated systems, without magnetism, are suitable for a model study.

The neighboring X-C bonds, as shown by the O- and S-decorated systems, possess the opposite distortions along the armchair edges. For the F and Cl adatoms, the first two attached with a carbon dimer lie above the ribbon plane while the following two are situated at the opposite position, and so on. Among these four kinds of adatoms, only the O-decorated system present a higher-energy planar structure, with two closed heptagon-pentagon edges, as observed in the N-decorated system (Figure 8.1(a)). Specially, the type-II oxygen passivation creates a short O-C bond length of 1.22 Å, with a high charge distribution by a strong covalent bonding. The bond lengths grow with the atomic number for the same group in the periodical table, such as, S-C>O-C and Cl-C>F-C (Table 8.1). With higher electronegativities, O and F can attract 1.73 e and 0.78 e from the neighboring carbon atom, respectively. Moreover, the cross section of the charge distribution, as shown in Figure 8.2(b), illustrates the covalent bondings in sulfur dimers and S-C bonds. In addition, it might have certain chemical bonding between two S adatoms associated with the neighboring dimers (green color).

The type-III systems, which consist of much bigger adatoms (X = Li, Na, K, Ca, Mg, Al, Si, and P), are characterized by the distorted and closed edge structures, as shown in Figure 8.1(c). The adatoms symmetrically stack up in the distorted pentagons, accompanied with the uniform carbon hexagons. All of them could transfer part of valence electrons to the bonding carbon atoms (Table 8.2), especially for Li, Na and K. Alkali adatoms have the lower electronegativity ($\sim 3/5$), compared with carbon atom. For example, the atomic interactions in the Li-C bonds are highly ionic, since most of valence changes are distributed closer to the edge C atoms (Figure 8.2(c)). The attractive interactions in the X-C bonds compete with the repulsive ones in the stacked adatoms; therefore, their competition determines the optimal X-C and X-X bond lengths. Among type-III systems, the P-decorated system has the shortest X-C and X-X bonds, indicating the significant atomic inter-

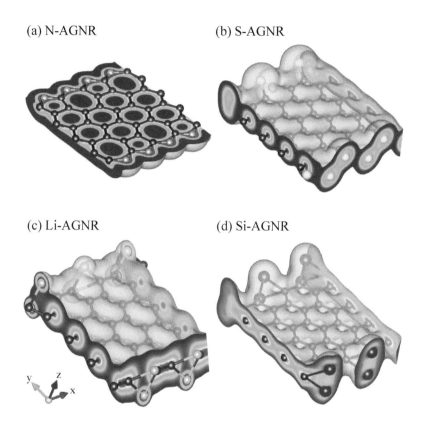

**FIGURE 8.2**
The spatial charge densities for (a) N-, (b) S-, (c) Li-, and (d) Si-decorated $N_A = 8$ armchair nanoribbons.

actions (Figure 8.2(d)). This directly reflects more valence charges in the P adatoms.

In general, the geometric structures of the edge-passivated graphene nanoribbons are dominated the complicated cooperation/competition of the atomic interactions among the C-C, X-C and X-X bonds. Whether the edge decorations could be achieved in the experimental growth strongly depend on adsorption energies. $E_{ads}$'s are, respectively, $-5.881$ eV, $-5.776$ eV, $-5.500$ eV, and $-4.795$ eV in the N-, O-, B-, and F-decorated systems, in which they are lower than that ($-4.359$ eV) of the H-decorated system frequently utilized in the experimental measurements. According to the detailed theoretical predictions, the first one exhibits the lowest adsorption energy among all the systems. On the other side, it might be difficult to synthesize the Na- and Ca-decorated systems with the higher adsorption energies of $-0.732$ eV and $-0.819$ eV, respectively. Up to now, some adatoms, which are as stable as a diatomic gas, have been applied to passivate the nanoribbon edge in a monatomic form, such as H [330], O [166], and Cl [333]. The open and distorted edge structures are confirmed by using the STM measurements and the X-ray single-crystal analyses. The experimental syntheses on the other adatom-decorated systems are useful in identifying the diverse geometric structures, covering the triangle-pentagon, pentagon-heptagon, and hexagon edges in planar profiles, and the edge-buckled and adatom-stacked-up configurations.

The strong orbital hybridizations in various chemical bonds, combined with the finite-size effects, can greatly diversify the electronic properties. A pristine $N_A = 8$ armchair nanoribbon, as shown in Figure 8.3(a), exhibits an indirect band gap of $E_g^i = 0.473$ eV. The highest occupied state and the lowest unoccupied state are situated at $k_x = 0$ and 1, respectively. The former and the latter are, respectively, dominated by the non-edge and edge carbon atoms (distinguished by the black circles, Figure 8.3(a)). Moreover, the dangling bonds at the edges are responsible for two pairs of valence and conduction bands (discussed earlier in Figure 3.2). The main features of band structures, atom dominance, energy dispersion and band gap, are dramatically changed after the significant edge passivation, as clearly indicated in Figures 8.3(b)–8.2(c). Apparently, the edge-carbon-dominated energy bands disappear in the presence of X-C bonds, i.e., the absence of dangling bonds will thoroughly modify the electronic structures.

The low-lying energy bands depend on the kinds of adatoms. The N-decorated system presents the linear and non-monotonous energy dispersions near the Fermi level, being associated with carbon atom and adatoms (inset in Figure 8.3(b)). Their band edge states at $k_x = 0$ determine a narrow direct gap of $E_g^d =0.04$ eV. Specially, the N adatoms make much contribution to four valence bands in the range of $-1.6$ eV$\leq E^{c,v} \leq -0.9$ eV, in which they are almost doubly degenerate and have weak energy dispersions (blue circles). As for the S-decorated system, there exists a pair of energy bands with parabolic dispersions close to $k_x = 0$ (Figure 8.3(c)). The smallest energy spacing is a middle direct gap of $E_g^d =0.395$ eV. The S adatoms dominate the shallowest

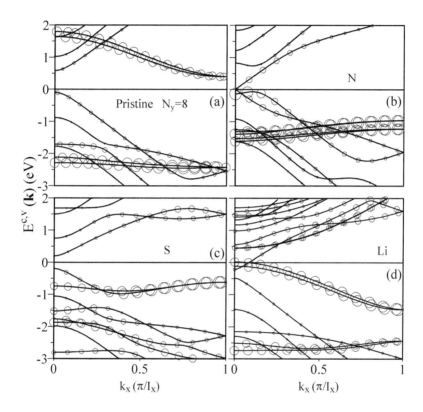

**FIGURE 8.3**
Energy bands of (a) pristine, (b) N-, (c) S-, and (d) Li-decorated $N_A = 8$
armchair nanoribbons, in which the contributions from the nonpassivated edge
carbon atoms or the adatoms are denoted by the open circles.

two valence bands (large yellow circles) except for the electronic states near the highest occupied one, Moreover, they contribute to some valence and conduction bands. Unusually, the Li adatoms could create the semiconductor-metal transitions; that is, the Li-decorated system might become a 1D metal (Figure 8.3(d)). The lowest conduction band overlaps with the two highest valence bands; the latter are associated with the Li adatoms (green circles). Some free electrons and holes, respectively, occupy the lowest conduction band and the highest valence one. Such a band overlap is also revealed in Na-, K-, Ca-, Mg-, and Be-passivated systems. This indicates that the quantum confinement is effectively suppressed by the charge redistributions due to the edge X-C and X-X bonds.

The distinct adatom decorations could dramatically alter the width dependence of energy gap, since the band-edge states near $E_F$ are sensitive to the edge passivation. For $N_A \geq 6$ pristine armchair nanoribbons, $E_g$ declines in the increase of width according to the $N_A = 3I$, $3I + 1$ and $3I + 2$ categories (Figure 8.4(a), as discussed earlier in Figure 3.4(a)). A linearly inverse relationship is absent, reflecting the strong competition between the edge dangling bond and the finite-size effect. Among three categories, energy gaps presnet a simple ordering: $E_g(3I + 1) > E_g(3I + 2) > E_g(3I)$ (red squares, blue dots, and green triangles). After the edge decoration, the X-C bonds and the quantum confinement account for energy gaps and might thoroughly change the specific behaviors. Three types of X-decorated systems, as shown in Figures 8.4(b)–8.4(d), have certain important differences in energy gaps, such as the irregular width dependence, the greatly reduced magnitudes, and the semiconductor-metal transition. Both N- and Li-decorated systems possess much smaller gaps (Figures 8.4(b) and 8.4(d)), compared with pristine ones. The former exhibit a non-monotonous $N_A$-dependence, and a severe fluctuation for the $N_A = 3I$ category (red squares in Figure 8.4(b)).The main reason is that N adatoms and C atoms co-dominate the energy dispersions nearest to $E_F$. Moreover, the significant Li-C bonding make the conduction and valence bands approach to $E_F$ for the $N_A = 3I$ and $3I + 1$ categories, or overlap for the $N_A = 3I + 2$ systems. The narrow gaps in the former two become vanishing when the widths are sufficiently large [$N_A \geq 15$ (red squares in Figure 8.4(d)) and $N_A \geq 22$ (blue dots)]. Apparently, the weakened finite-size effect leads to the semiconductor-metal transition. On the contrary, the S-decorated systems behave similar to pristine ones (Figure 8.4(c)), since the adatoms are almost independent of the highest occupied state and the lowest unoccupied state (Figure 8.3(c)). The adatom-enriched energy gaps could be examined by the STS experiments. The tunable energy gaps and free carrier densities are very useful for potential applications, e.g., FETs and light-emitting diodes.

The atom- and orbital-decomposed DOSs, as illustrated in Figures 8.5(a)–8.5(d), can comprehend the decoration effects related to the C-C, X-C and X-X bondings. There are a lot of asymmetric peaks and part of symmetric peaks (red circles) arising from the parabolic and weakly dispersive bands, respectively. The $N_A = 8$ pristine armchair nanoribbon presents a pair of asym-

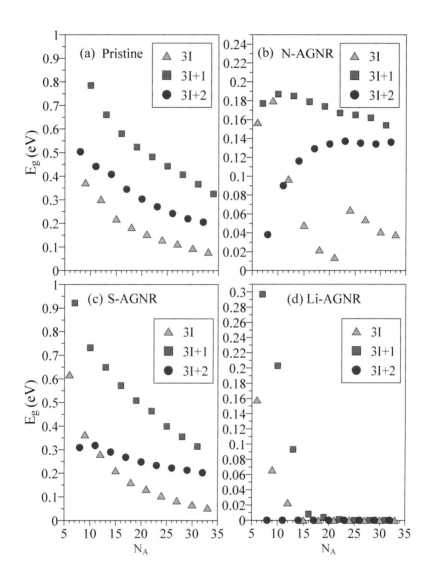

**FIGURE 8.4**
The width-dependent energy gaps for the (a) pristine, (b) N-, (c) S-, and (d)
Li-decorated armchair nanoribbons.

metric peaks across the Fermi level at $E = \pm0.24$ eV, in which the valence and conduction band-edge states are exclusively contributed by the $2p_z$ orbitals (red curves in Figure 8.5(a)) and the $(2s, 2p_x, 2p_y)$ orbitals (blue and green curves), respectively. The $2p_z$ orbitals (the $\pi$ bondings) make important contributions in the range of $-3$ eV$\leq E \leq 3$ eV, while the other orbitals (the $\sigma$ bondings) dominate those of $E \leq -2.2$ eV and 0.2 eV$\leq E \leq 1.7$ eV. The $\sigma$-orbital contribution to the low-lying conduction states mainly arises from the edge dangling bonds. This is fully replaced by the $2p_z$-orbital contribution in the presence of edge passivation (red curves in Figures 8.5(b)–8.5(d)). Furthermore, the $2s$-orbital DOS is largely reduced at $E \geq -3$ eV.

The adatoms have drastically changed the energy, number, height of the special structure in DOS. Their DOS corresponds to the difference between the total DOS and that of carbon atoms. Specifically, the N-decorated system exhibits the adatom-dominated shoulder structure near $E_F$ and six prominent peaks in the range of $-1.6$ eV$\leq E \leq -0.9$ eV (Figure 8.5(b)). Such peaks are combined with the significant $(2p_x, 2p_y)$ contributions, indicating the strong interactions among the N adatoms and the dominant carbon orbitals in a planar structure. Specially, the S-enriched peaks appear in the wide ranges of $E \leq -0.65$ eV and $E \geq 1.3$ eV, being accompanied with the $2p_z$-orbital ones. The large energy width clearly confirms the existence of S-S bonds in the distorted edges. The merged peaks arise from the non-planar S-C bonds covering valence charges of adatoms and carbon $2p_z$ orbitals. As to the Li-decorated system, the one valence electron in each adatom only makes a partial contribution to DOS, especially for the range of $-1.5$ eV$\leq E \leq 0$ eV. A finite DOS is revealed at $E_F$, in which the metallic property is ascribed to the critical chemical bonding between Li adatom and carbon atom.

## 8.2    Decoration- and curvature-enriched essential properties

The edge passivation in the unzipping process of carbon nanotube, as shown in Figures 8.6(a) and 8.6(b), is investigated by the first-principles method. The zigzag lines along the nanotube axis are relatively easily to be separated, compared with the armchair ones. Furthermore, their decorations can create more types of geometric structures and thus diversify essential properties. The adatoms, with an atomic number less than 30, are included in the investigations. An armchair carbon nanotube, corresponding to a zigzag graphene nanorbbon, is chosen for a model study. When it is being cut longitudinally, the odd- and even-$N_Z$ zigzag nanoribbons result. An odd-width nanoribbon is more stable, so this work is focused on its decoration effects. By the detailed calculations, the optimal geometric structures cover three types: (I) a zipper-line decorated carbon nanotube (Figure 8.6(c)), (II) an edge-decorated curved

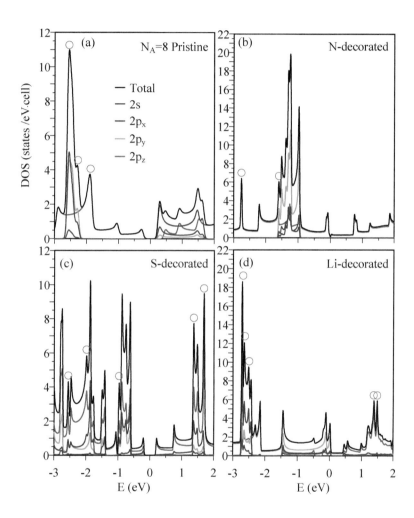

**FIGURE 8.5**
DOSs of the (a) pristine, (b) N-, (c) S- and Li-decorated $N_A = 8$ armchair
nanoribbons. Also shown are the orbital-projected components due to the
carbon atoms. The symmetric peaks are denoted by the red circles.

nanoribbon (Figure 8.6(d)) and (III) a flat nanoribbon with edge passivation (Figure 8.6(e)). The formation of each geometric structure is first discussed through a detailed analysis of its total energy. The different electronic configurations of decorating adatoms bring about various types of edge-carbon-adatom interactions. These are categorized into three types, being identified from the charge transfer, C-C bond length, etc. Moreover, the adatom distribution is very useful for further understanding the electronic properties.

The arc angle in range of $0° \leq \phi \leq 360°$, as defined in Section 4.1, can characterize a curved/planar nanoribbon, and a unzipped/zipped nanotube. In order to determine an optimal geometric structure, the $\theta$-dependent total energy ($\Delta E_{tot}$), being measured from that of a flat nanoribbon, is calculated for each type of decoration configuration. $\Delta E_{tot}$, as shown in Figure 8.7(a), presents three types of adatom-dominated arc angle dependences. The critical arc angle, corresponding to the local maximum of $\Delta E_{tot}$, plays a critical in determining a stable configuration. When $\phi_c$ is larger than the initial arc angle, the zipper-line-decorated nanotube (type I; red circles) and highly curved ribbon (type II; blue squares) are formed. On the contrary, the resulting structure is a planar nanoribbon with edge passivation. Moreover, all type-III structures are planar in the absence of $\phi_c$ (green triangles). The $\phi$ dependence of $\Delta E_{tot}$ is mainly attributed to the mechanical strain and the edge-edge atomic interactions. These two effects cooperate in deciding the resulting decorated structures for type III. However, they become a strong competition for type I and type II. Specially, the $\phi$-dependent energies in type-II systems are well fitted as [57]

$$\Delta E_t \simeq [3.76 \times 10^{-6} \times \exp(\frac{13.34}{N_y})]\theta^2 + 2\sum_{\alpha=1}^{2} \frac{(\overrightarrow{p_{l,\alpha}} \cdot \overrightarrow{p_{r,\alpha}}) - 3(\overrightarrow{p_{l,\alpha}} \cdot \hat{n})(\overrightarrow{p_{r,\alpha}} \cdot \hat{n})}{4\pi\epsilon_0 r_\alpha^2},$$

$$(8.1)$$

The first term in Equation (8.1) is the mechanical strain energy inversely proportional to $\phi$ (radius), as discussed earlier in Section 4.1. Furthermore, the second term represents the dipole-dipole interactions from the edge-dependent electric dipole moments [141]. $\epsilon_0$ is the free-space dielectric constant, $\overrightarrow{p_{l,\alpha}}$ ($\overrightarrow{p_{r,\alpha}}$) the dipole moment on the left-hand (right-hand) side, $\hat{n}$ the unit vector along two moments, and $r_\alpha$ the distance between them. $\alpha=1$ and 2, respectively, represent the nearest and next-nearest neighbors. The factor of 2 indicates a dipole interacting with one dipole on each side. The detailed discussions will be provided later.

Different decorating adatoms (e.g., B, Be, H, and C) lead to the dramatic changes of the total energies and thus decide the   optimal geometric structures. According to the arc angle dependence, $\Delta E_{tot}$ could be divided into two zones: (I) $\phi < \phi_c$ and (II) $\phi > \phi_c$). This is clearly shown in Figure 8.7(a) for the $N_Z = 7$ system. In zone I, it is almost identical for all types of edge-decorated curved systems, in which the $\phi$-square dependence agrees with the first term of Equation (8.1), as revealed in carbon nanotubes [291]. The increase of the total energy directly reflects the enhanced mechan-

(a) Armchair carbon nanotube     (b) Unzipped nanotube

(c) Type I             (d) Type II

(e) Type III

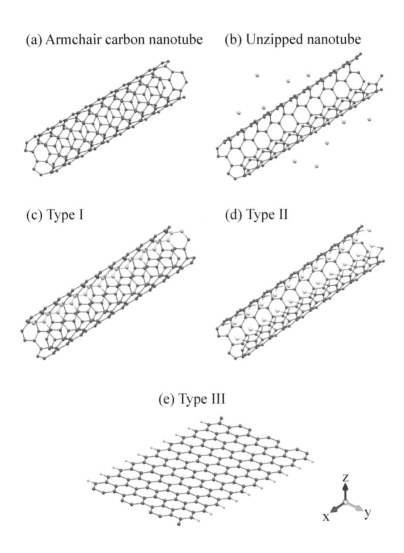

**FIGURE 8.6**
Geometric structures: (a) A single-walled armchair carbon nanotube as a starting material, (b) an unzipped nanotube in an environment of decorating adatoms, (c) type-I, (d) type-II, and (e) type-III adatom-decorated nanotube/nanoribbon-related systems.

ical strain. Zone II refers to a region at sufficiently large $\phi$'s. Specially, the adatom-dependent total energies start to behave differently and diverge from one another. The edge-edge interactions need to be taken into account when the inter-edge distance is getting smaller. In general, various edge adatoms result in three different types of edge interactions: strong covalent bonds, significant dipole-dipole interactions, and dominant repulsive forces. For type-I boron decoration, its energy declines quickly as the arc angle exceeds the critical one. The slopes in the cases of boron (red circles) and carbon (black crosses) atoms are similar, clearly indicating that the resulting configuration is ascribed to the very strong $sp^2$ covalent bonds. The bonding energy could be fitted by the linear superposition of $1/d$ and $1/d^2$, where $d$ is the distance between two edges. Concerning type-II structure, the quasi-stable configuration in beryllium case (blue squares in Figure 8.7(a)) is an open curved nanoribbon with large charge transfers between Be adatoms and C atoms (discussed later), being characterized by the dipole interaction model. The nearest and next-nearest dipole-dipole interactions in Equation (8.1) match the non-monotonous $\phi$-dependence of $\Delta E_{tot}$ well. In comparison with the boron and the carbon cases, the decrease in total energy happens at a smaller critical angle, mainly owing to the longer interactive distance of non-bonding forces. The hydrogen-decorated nanoribbon belongs to type III, which is very different from types I and II. As the arc angle gradually grows, the repulsive interactions become strong and thus enhance the total energy (green triangles in Figure 8.7(a)). The edge adatoms are in saturated states so that no chemical bonding forms between two edges, and the charge transfer between H and C atoms is too small in creating the effective dipole-dipole interactions. As a result, the quasi-stable structures are absent; that is, only a flat nanoribbon is allowed.

The formation of type-II structures could be understood from the energy variation in zone II (blue squares). With the increasing arc angle, the reduction in the total energy is attributed to the shorter distance and the approximately anti-parallel arrangement for the two-side dipoles (most contributions from the first part of the second term in Equation (8.1)). The enhanced attractive interactions lead to a local minimum energy at a specific arc angle. Furthermore, the energy difference between the curved nanoribbon (the dashed black curve) and the Be-decorated system represents that created by the dipole interactions. For Be and Mg adatom decorations, such energy differences are, respectively, 1.9 eV and 1.1 eV in the first-principles calculations, being comparable to 2.1 eV and 1.2 eV by calculating the second term of Equation (8.1). This further illustrates that the dipole interaction model can account for the optimal type-II configurations. The type-II decorating adatoms exhibit the similar behavior in the highly curved nanoribbons, while various electric dipole moments result in different curvatures of the quasi-stable structures. In addition, the total energy will grow in the further increase of $\phi$, e.g., the extra repulsive interaction energy at $\phi = 2\pi$.

For type-I and type-II configurations, the critical interaction distance and

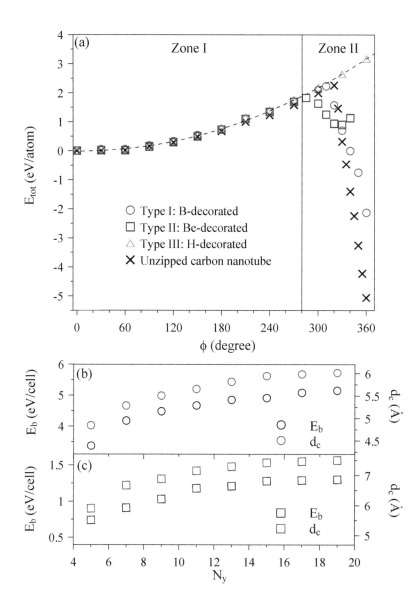

**FIGURE 8.7**
(a) The arc-angle-dependent total energies, measured from those at $\phi = 0°$, for the B-, Be-, and H-decorated $N_Z - 7$ armchair nanoribbons, as well as an unzipped carbon nanotube. The width dependence of the critical interaction distance and the energy barrier in (b) type-I and (c) type-II adatom-decorated systems.

the energy barrier ($E_b$) in preventing the production of flat structures strongly depend on the ribbon width, as indicated in Figures 8.7(b) and 8.7(c), respectively. Because of the reduced mechanical strains, the wider ribbons only require the relatively weak edge-atom interactions to prevent flat ribbons from occurring. This is responsible for the straightforward increasing relationship between $N_Z$ and $d_c$ (red solid circles in Figure 8.7(b) and blue solid squares in Figure 8.7(c)). $E_b$, the total energy difference between the stable (quasi-stable) structure and the critical one, for boron (beryllim) will gradually rise in the increase of $N_Z$. Furthermore, the tendency of $E_b$ and $d_c$ levels off for $N_Z \geq 15$. These are mainly determined by the competition between the bending energy and the attractive edge-atom interactions. According to Figures 8.7(a) and 8.7(b), the type-I and type-II configurations, with sufficiently large widths, could form spontaneously when the interaction distance is less than 7 Å.

The zipped/curved/planar adatom-decorated zigzag nanoribbons present very rich characteristics, such as charge transfers, X-X and X-C bond lengths, magnetic moments and metallic/semiconducting behaviors listed in Table 8.2. For boron and nitrogen adatoms, classified as type I, the strong covalent bonds are created in X-X and X-C bonds. To create type-I structures, it is essential that the number of occupied valence electrons, X-X distance, X-C distance and electronegativity are all close to those of carbon. A boron atom possesses three valence electrons. Two of them are assumed to form the two B-B single bonds and thus make a zigzag adatom chain, while the third one causes the B-C single bond to zip the curved ribbon into a cylindrical nanotube. However, a nitrogen adatom can generate two N-N single bonds and one N-C single bond, in which two unshared electrons are left for making up a full octet. This bonding configuration induces a serious deformation of a zipped nanotube. The zigzag nitrogen atoms, which protrude from the non-uniform cylindrical surface, are responsible for the anti-ferromagnetism of the edge carbon atoms ($\sim 0.13 \ \mu_B$)), as revealed in deformed boron nitride nanotubes [392]. Except for B and N, other non-metal elements, e.g., O, S, F, Cl, Si and P, can bond only with carbon atoms, but not form a tubular structure in the absence of the three above-mentioned essential conditions. As a result, these elements exhibit the edge passivation in flat zigzag nanoribbons.

It is well known that transition metals play critical catalysts in synthesizing carbon nanotubes [140, 337, 62]. According to the theoretical predictions, seven of the transition metals in the fourth row of the periodic table can exhibit type-I nanotube structures (Table 8.2). Such elements have many $3d$ orbitals and provide occupied/unoccupied states, leading to the X-X single bond. While the X-X bond lengths in transition metals are larger than those of boron and nitrogen, they still remain stable under the moderate deformations of tubular structures. Specifically, transition metals can give rise to magnetic moments on the edge carbon atoms except that non-ferrous metal copper possesses a nearly zero magnetic moment. The highly curved type-II structure is only revealed in Ti adatoms, directly reflecting that they have only two

**TABLE 8.2**
The $N_Z = 7$ zigzag graphene nanoribbon decorated by adatoms with an atomic number less than 30, revealing the type-I, type-II, or type-III structures (the tubular, highly curved, or flat structures), charge transfers, X-C bond lengths, X-X bond lengths, magnetic moments, spin configurations, and semiconducting/conducting behaviors.

| Type | Atom | Charge Transfer* (e) | A-C Distance (Å) | A-A Distance (Å) | $(\mu_B)$ C/Atom | AFM/ FM | (Semi-) conductor |
|------|------|------|------|------|------|------|------|
| | B | 0.62 | 1.514 | 1.695 | 0.00/0.00 | | M |
| I | C | 0.00 | 1.424 | 1.424 | 0.00/0.00 | | M |
| | N | -0.18 | 1.427 | 1.522 | 0.13/0.01 | AFM | S |
| | Be | 1.30 | 1.823 | 2.899 | 0.00/0.00 | | M |
| II | Mg | 1.10 | 2.146 | 3.124 | 0.00/0.00 | | M |
| | Al | 0.55 | 2.058 | 2.899 | 0.00/0.00 | | M |
| | H | 0.41 | 1.092 | | 0.15/0.00 | AFM | S |
| | Li | 0.81 | 1.911 | | 0.14/0.00 | AFM | S |
| | Na | 0.49 | 2.385 | | 0.00/0.00 | | M |
| | K | 0.35 | 2.656 | | 0.00/0.00 | | M |
| | Ca | 0.71 | 2.290 | | 0.00/0.00 | | M |
| III | Si | 0.92 | 1.880 | | 0.12/0.02 | AFM | M |
| | P | 0.58 | 1.761 | | 0.00/0.00 | | M |
| | O | 0.59 | 1.255 | | 0.00/0.00 | | M |
| | S | 0.12 | 1.719 | | 0.00/0.00 | | M |
| | F | -0.43 | 1.349 | | 0.14/0.03 | AFM | S |
| | Cl | -0.42 | 1.734 | | 0.00/0.00 | | M |
| | V | 0.52 | 2.119 | 2.422 | 0.03/0.70 | AFM | M |
| I | Cr | 0.54 | 2.110 | 2.542 | 0.14/2.98 | AFM | S |
| | Mn | 0.46 | 2.094 | 2.557 | 0.12/3.21 | FM | M |
| | Fe | 0.25 | 1.895 | 2.447 | 0.16/2.32 | FM | M |
| | Co | 0.23 | 1.841 | 2.299 | 0.00/0.51 | AFM | M |
| | Ni | 0.18 | 1.844 | 2.428 | 0.00/0.25 | FM | M |
| | Cu | 0.18 | 1.934 | 2.520 | 0.00/0.00 | | M |
| II | Ti | 0.58 | 2.107 | 2.848 | 0.13/0.15 | FM | M |
| III | Sc | 0.45 | 2.213 | | 0.08/0.73 | FM | M |
| | Zn | 0.52 | 2.267 | | 0.19/0.03 | AFM | M |

*The plus and minus signs, respectively, denote attracting and repelling electrons.

occupied valence electrons, the high charge transfer and the long X-X bond (also see next paragraph). Concerning Zn and Sc, electrons fully occupy the 3*d* orbitals of the former and one 3*d* electron appears in the valence state of the latter. These two kinds of adatoms could bond with carbon atoms, but it is impossible to hybridize with each other. Furthermore, their charge transfers are too low to create any dipole moments. As a result, the Zn- and Sc-decorated zigzag nanoribbons present the type-III planar structures.

For a zigzag nanoribbon decorated with metal elements, the high charge transfer between adatoms and carbons could create the significant electric dipoles on the edges and thus stabilize the type-II curved structures. The self-consistent calculations reveal the much longer X-X distances, compared with those in type I. The metal-decorated systems do not present the deformed/non-uniform cylindrical structures, since the long interatomic distances make it impossible for the metal atoms to share electrons in single bonds. Both Be- and Mg-decorated type-II structures mainly come from the appropriate X-X distance and the relatively high electron transfer. Such essential conditions cause the electronic dipole effects to become dominant. An Al-decorated structure is regarded as a special case, in which the relatively low charge transfer could induce sufficient dipole moments to achieve the highly curved structure. The Al adatoms have three valence electrons, while the very large differences among the X-X, X-C, and C-C bond lengths could prevent the formation of type-I nanotube structures. The Ti adatoms, which possess the highest electron transfer among the transition metals, show the similar behavior. They give rise to ferromagnetic moments of edge carbon atoms, being absent in other type-II structures. On the other side, the only one valence electron in alkali metal adatoms cannot provide enough charge transfer for generating type-II systems. The decoration by other metal elements such as Ca does not result in type-II structures when the X-C bond length is too long. This could be attributed to the lengthy bond that behaves as electric monopoles, leading to a repulsive force between two-side edges. Li, Na, K, and Ca, for instance, form stable structures of flat adatom-decorated nanoribbons.

The electronic properties of nanoribbon-related systems are greatly diversified by the adatom decorations and the curvature variations. As for pristine systems, an armchair nanotube has two linear bands intersecting at the Fermi level ((4,4) in Figure 8.8(a)), while a flat zigzag nanoribbon has a pair of partial flat bands with a indirect gap (Figure 3.8 in Chapter 3). These two kinds of energy bands are, respectively, dominated by all carbons and edge ones. The low-energy features dramatically change with the various adatoms. A B-decorated cylindrical nanotube, as clearly indicated in Figure 8.3(b), exhibits two unique features: three energy bands crossing the Fermi level, and a pair of partial flat bands below it. The first linear energy band intersecting with $E_F$ near $k_x = 1/2$ mainly originates from the B-B bond in the adatom zigzag chain, with the dominance represented by the radius of circle. The other two nonlinear bands are closely related to B-C bonds and most C-C bonds.

The carrier transport along the edge B-B and B-C bonds is deduced to be quite important. On the other hand, a pair of partial flat bands attributed to the edge C-C bonds has no energy spacing at the zone boundary, reflecting the absence of anti-ferromagnetism at both edges (different from Figure 8.8(d)). Specifically, the partial flat bands and one linear band are absent in other nanotube-related systems where boron atoms are uniformly distributed, e.g., $BC_3$ [157] and $BC_2N$ nanotubes [2]. This difference further illustrates the geometry-enriched electronic properties. In addition, a N-decorated nanotube, with a protruding surface, is a direct-gap semiconductor of $E_g = 0.2$ eV (Table 8.2).

The low-lying energy bands of a highly curved nanoribbon present the composite features related to a H-terminated flat nanoribbon (Figure 8.3(d)) and a carbon nanotube (Figure 8.3(a)). The Be-decorated curved nanoribbon, as shown in Figure 8.3(c), has a pair of partially flat bands below $E_F$ mainly arising from the edge atoms, and two linear bands intersecting at $E_F$. These clearly indicate that the Be-decorated nanoribbon is a 1D metal, whereas the H-terminated one is a semiconductor with two separated flat bands. One of the linear bands is dominated by all the carbon atoms, and the other is partially contributed by the Be-C bonds (green circles). As a result, electrons can be transported either through uniformly through the carbon atoms or specifically through the decorating adatoms. These characteristics are also revealed in larger-width systems. Other type-II systems, such as Mg- and Al-decorated ones, exhibit the similar low-energy behaviors. Although the origins of linear bands might be somewhat different, the type-II systems possess the almost same peculiar feature as the armchair carbon nanotubes: the metal-like electrical conductivity required for use as 1D nanowires, being a promising feature to apply in future nanoelectronics.

The low-energy DOSs in edge-decorated curved zigzag nanoribbons, as shown in Figures 8.9(a)–8.9(d), could exhibit a plateau across $E_F$, two prominent peaks, and several asymmetric peaks in the square-root form. The plateau DOS (Figures 8.9(a) and 8.9(c)) is due to a pair of linearly intersecting bands in an armchair nanotube (Figure 8.8(a)) and a Be-decorated nanoribbon (Figure 8.8(c)). The negligible $E$-dependence is absent in the metallic B-decorated nanotubes (Figure 8.9(b)) because of the two nonlinear energy bands (Figure 8.8(b)). This system might have the highest DOS at $E_F$ among the 1D nanostructures, so it is expected to present a very high electrical conductivity. The two prominent peaks exist in the adatom-decorated systems (red triangles in Figures 8.9(b)–8.9(d)), clearly indicating the important contributions of edge carbon atoms (red curves). They are located separately above and below $E_F$ for the H-terminated flat nanoribbon, while the B- and Be-decorated curved/tubular systems show two neighboring structures beneath $E_F$. Specifically, both B- and H-decorated systems possess an extra peak (the blue circle in Figures 8.9(b) and 8.9(d)) near the prominent peaks. This special structure, respectively, arises from the $k_x = 0$ state of the $B$-dependent nonlinear band

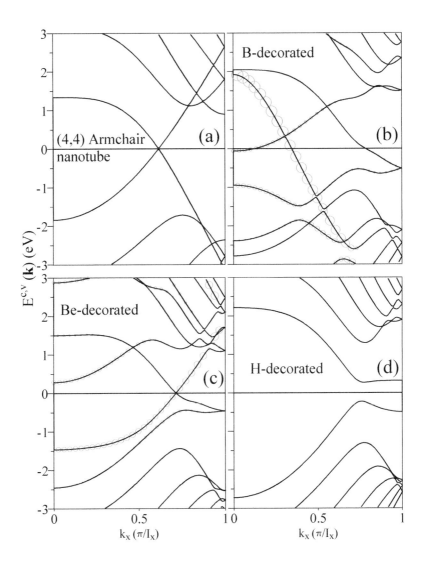

**FIGURE 8.8**
Band structures for (a) (4,4) armchair nanotube, (b) B-, (c) Be-, and (d) H-decorated $N_Z = 7$ systems, in which the adatom contributions are represented by the circle radii.

(blue curve; Figure 8.8(b)) in the former and the $k_x \sim 3/4$ state of the partial flat band in the latter (Figure 8.8(d)).

The metallic/semiconducting behavior of an adatom-decorated zigzag depends on the magnitude of DOS at $E_F$, as listed in Table 8.2. Most decorating atoms could provide free carriers and thus enhance $DOS(E_F)$, i.e., such decorated systems are conductors. Compared to metallic armchair carbon nanotube, they possess a much higher $DOS(E_F)$ except for Be- and Mg-decorated systems with a comparable value. When the anti-ferromagnetic configuration induced by zigzag edges comes to exist, the partial flat bands near $E_F$ become splitting and create a vanishing $DOS(E_F)$. For example, the H-, Li-, and F-terminated zigzag nanoribbons belong to semiconductors. However, the Si-decorated system is a metal even if it owns an AFM configuration. The main reason is that the Si adatoms, with more valence electrons, can dominate the the valence electrons across $E_F$. As to the transition-metal-related systems, they tend to maintain the intrinsic properties of the decorating atoms, and hence are conductors. The only exception is a tiny-gap Cr-decorated system. This gap might disappear in wider systems The feature-rich DOSs near $E_F$ could be examined by the STS measurements to identify the decoration and curvature effects, such as the adatom-dependent $DOS(E_F)$ and peaks, and the edge-C-dominated peak energies.

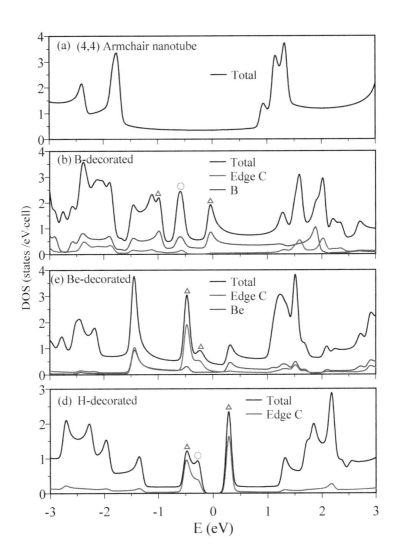

**FIGURE 8.9**
DOSs and local DOSs due to adatoms and edge carbons for (a) (4,4) arm-chair carbon nanotube, (b) B-, (c) Be-, and (d) H-decorated $N_Z = 7$ zigzag nanoribbons.

# 9

## Alkali-adsorbed graphene nanoribbons

In addition to zipping, folding, stacking, scrolling, and edge decoration, the surface adsorption is a very efficient method in diversifying the essential properties of graphene nanoribbons. Alkali adatoms are very suitable for a model study, since only the outmost s-orbital plays a critical role in the active chemical environment. The single-orbital hybridization of the X-C bond is relatively easy in understanding the alkali-induced dramatic changes. The alkali-created chemical modifications on the carbon-related systems with $sp^2$ bondings have been extensively studied both experimentally and theoretically, such as graphite [360], fullerenes [267], carbon nanotubes [39], few-layer graphenes [128], and graphene nanoribbons [230]. Such systems after alkali intercalation/adsorption, such systems can present very high conduction electron densities and thus electrical conductivities. The n-type superconductivities are observed in graphite interaction compounds; alkali-doped $C_{60}$ fullerides [115], and carbon nanotube bundles [22]. Specially, Li-based graphene compounds could serve as anode of high-performance batteries, in which they are very durable within the frequent charging processes. Lithium is smallest among all the alkalis, so Li-adsorbed graphene nanoribbons are expected to be most stable and present high concentrations in experimental syntheses. They are chosen to illustrate the alkalization effects.

The chemical modifications on the essential properties of graphene nanoribbons could be easily achieved by the alkali surface adsorptions under the active environment, especially for the adatom-induced free carrier density and the diversified spin arrangements. The significant orbital hybridizations in X-C, C-C, and X-X bonds need to be examined in detail from the atom-dominated energy bands, the spatial charge distributions, and the atom- and orbital-projected DOSs. The geometric features of alkali-adsorbed graphene nanoribbons, the binding energy, optimal position, C-C and C-X bond lengths, and adatom height, are included in the first-principles calculations. The effects of alkali adatoms on conduction and valence bands are investigated for various adsorption configurations, covering the variations in the kind of adatom, concentration, relative position, single or double sides, edge structure and width. The systematic calculations and analyses are available in determining the specific/simple relation between the linear free carrier density and the adatom concentration. Specially, the position and concentration of alkali adatoms on zigzag nanoribbon will be tuned to explore the dramatic changes in the partially flat bands and the spatial spin density, such as the spin-split edge-localized bands and the transformation from anti-ferromagnetism to non-magnetism/ferromagnetism.

## 9.1   The alkali-created conduction electrons

Graphene nanoribbons, with open edges passivated by hydrogen atoms, are adsorbed by alkali atoms on surfaces. The geometric, electronic and magnetic properties are investigated under the different alkali adatoms, concentrations, distributions, widths, and edge structures. Both armchair and zigzag systems behave similarly except for magnetic configurations, so the former are suitable for a model study in the alkali-induced free carriers. The optimal adatom position, as shown in Figure 9.1(a) for $N_A = 10$ armchair nanoribbon, is the hollow site, but not the bridge and top sites. The similar position in 2D graphene has been confirmed by the low-energy electron microscopy [347]. The binding energy of each adatom, the  reduced energy arising from alkali adsorption, is defined as $E_b = [E_{sys} - E_{gra} - nE_A]/n$, where $n$ is the number of adatoms; $E_{sys}$, $E_{gra}$ and $E_A$ are, respectively, the total energies of the alkali-adsorbed system, graphene nanoribbon, and isolated alkali atoms. Li and Na adatoms, respectively, possess the highest and lowest binding energies ($-1.14$ eV and $-0.55$ eV in Table 9.1(a)). It is relatively easy to produce Li adsorption in experimental syntheses among all alkali adatoms. As to the adatom height (the A-C bond length), it varies from 1.80 Å to 3.03 Å (2.31 Å to 3.35 Å) as Li→Cs. Graphene nanoribbon remains the planar structure after alkalization for any concentrations and distributions. This clearly indicates that the $\sigma$ bonding due to $(2s, 2p_x, 2p_y)$ orbitals of carbon atoms is almost unchanged by alkali adsorption. However, the C-C bond lengths are sensitive to the positions of carbon atoms, but not those of alkali adatoms. In addition, the total ground state energy only exhibits a slight variation (about several meV's) for the distinct adatom positions.

The alkali-atom adsorption can dramatically alter band structures of graphene nanoribbons, especially for the blue shift of the Fermi level (the creation of free electrons in conduction bands). We first discuss the single-side adsorption with one alkali  adatom $[(1)_s]$ situated at the center of edge hexagon (Figure 9.1(a)). An energy spacing ($\sim 1.18$ eV) between the valence and conduction bands almost keeps the same after the alkali adsorption, as indicated from those in pristine (Figure 9.2(a)) and alkali-doped systems (Figures 9.2(b)–9.2(f)). Furthermore, the valence bands are dominated by carbon atoms and hardly depend on alkali adatoms (very small red blue circles), further illustrating the unchanged $\sigma$ bonding and the slightly distorted $\pi$ bonding. On the other hand, the asymmetry of energy bands about $E_F$ becomes more obvious after alkali adsorption. Specially, some conduction bands are fully/partially dominated by alkali adatoms, e.g., the first/second/third con-

**TABLE 9.1**

(a) Binding energies, C-C and C-X bond lengths, alkali heights, and free electron densities of $N_A = 10$ alkali-adsorbed graphene nanoribbons with one adatom in a unit cell; (b) free electron densities for various distributions and concentrations of adatoms in $N_A = 12$ and $N_z = 8$ Li-adsorbed graphene nanoribbons.

(a)

| $N_A = 10$ | $E_b$ (eV) | C-C (Å) | C-X (Å) | Heights (Å) | $\lambda$/Adatom (e) |
|---|---|---|---|---|---|
| Li | -1.14 | 1.381 | 2.24 | 1.76 | 0.997 |
| Na | -0.55 | 1.372 | 2.29 | 1.83 | 1.015 |
| K | -0.87 | 1.369 | 3.22 | 2.91 | 0.995 |
| Rb | -0.75 | 1.370 | 3.34 | 3.05 | 1.011 |
| Cs | -0.72 | 1.368 | 3.38 | 3.09 | 0.995 |

(b)

| Graphene Nanoribbon | Configurations | $\lambda$ ($10^7$ $e/cm$) | $\lambda$/Adatom Concentration (e) | Bader Charge Transfer (e) |
|---|---|---|---|---|
| | $(1)_{single}$; Li | 2.31 | 0.995 | 0.75 |
| | $(1)_{single}$; K | 2.31 | 0.995 | 0.45 |
| | $(1)_{single}$; Cs | 2.31 | 0.995 | 0.54 |
| | $(3,7)_{single}$; Li | 4.67 | 1.008 | 0.69 |
| | $(3,7)_{double}$ | 4.62 | 0.989 | 0.75 |
| Armchair | $(1,10)_{single}$ | 4.68 | 1.008 | 0.74 |
| $N_A = 12$ | $(1,10)_{double}$ | 4.67 | 1.008 | 0.74 |
| | $(1,4,7,10)_{double}$ | 9.24 | 0.994 | 0.71 |
| | $(1,2,5,6,9,10)_{double}$ | 13.96 | 1.007 | 0.70 |
| | $(1,3,5,7,9,2,4,6,8,10)_{double}$ | 23.03 | 0.988 | 0.68 |
| | $(1{\rightarrow}10,1{\rightarrow}10)_{double}$ | 46.36 | 0.995 | 0.68 |
| | $(1)_{single}$ | 2.02 | 1.001 | 0.88 |
| Zigzag | $(7)_{single}$ | 2.03 | 1.009 | 0.89 |
| $N_Z = 8$ | $(1,13)_{single}$ | 4.10 | 1.015 | 0.88 |
| | $(1,13)_{double}$ | 4.02 | 1.001 | 0.88 |
| | $(1,7,13)_{single}$ | 6.13 | 1.011 | 0.87 |
| | $(1,7,13)_{double}$ | 6.08 | 1.002 | 0.88 |

(a) Armchair GNR    $N_A$=10

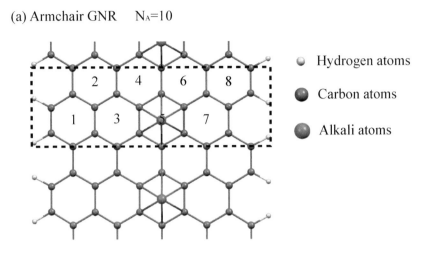

(b) Zigzag GNR    $N_Z$=10

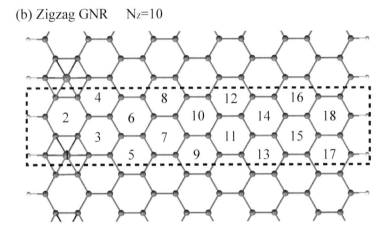

**FIGURE 9.1**
Geometric structures of alkali-adsorbed graphene nanoribbons: (a) $N_A = 10$ armchair and (b) $N_Z = 10$ zigzag systems with unit cells enclosed by the dashed rectangles. The gray, white, and purple balls, respectively, represent carbon, hydrogen, and alkali atoms.

duction band measured from $E_F$. They might have unimportant differences among the various alkali systems. These clearly indicate that the significant interactions between the neighboring alkali adatoms is able to generate energy bands. Most important of all, the Fermi level presents a blue shift; that is, it shifts from the middle of energy gap to the conduction bands. $E_F$ is determined by the alkali-induced conduction electrons. The Fermi momentum $(k_F)$ can characterize the 1D free carrier density $(\lambda_e)$ by a simple formula: $\lambda_e = 2k_F/\pi$. The dependence of $\lambda_e$ on the kind, distribution and concentration of alkali adatoms will be fully clarified in Table 9.1(b). In addition, energy bands are hardly affected by the various positions under the single-adatom adsorption.

Band structures of alkali-adsorbed systems strongly depend on the concentrations, relative positions, single- and double-side adsorptions, and edge structures (Section 9.2). The number of the alkali-dependent conduction bands is greatly enhanced in the increase of adatom concentration, as clearly indicated in Figures 9.3(a)–9.3(f) for various Li adsorptions (the blue circles). More conduction bands become occupied, indicating the rapid generation of free carriers during the alkalization process. Specially, each conduction band is closely related to the alkali adatoms under the full surface adsorption $[(1\text{-}8)_s$ in Figure 9.3(d) and $(1\text{-}16)_d$ in Figure 9.3(f)]. In general, the dependence of valence bands on adatom concentrations is weak. The low-lying conduction bands near $E_F$ are sensitive to the relative positions, e.g., $(4\,\text{Li})_s$ and $(4,\,\text{Li})_s$ in Figures 9.3(b) and 9.3(c), respectively. Also, they are drastically changed by the single- and double-side adsorptions $[(8\,\text{Li})_s$ in Figure 9.3(d) and $(8\,\text{Li})_d$ in Figure 9.3(e)]. By the detailed examinations, even although the Fermi momenta in the occupied conduction bands present obvious changes, their sum keeps the same under a specific Li concentration. This is irrespective of the kind of alkali adatoms.

After systematic calculation and analysis, the total conduction electron density below $E_F$ is linearly proportional to the alkali concentration, as clearly revealed in Table 9.1(b). $\lambda_e$ hardly depends on the adatom kind, relative position, single- or double-side adsorption, ribbon width, and edge structure; that is, the ratio between the free carrier density and the adatom concentration is very close to one under various adsorption configurations. This means that each alkali adatom contributes the outmost s-orbital electron as conduction carrier through the significant X-C, C-C ($\pi$ bonding), and X-X bondings ($\rho$ in Figure 8.2). As to $N_A = 10$ alkali-adsorbed systems, $\lambda_e$ is $\sim 2.34 \times 10^7$ e/cm for a single-adatom adsorption, and it could reach $\sim 3.18 \times 10^8$ e/cm in the double-side full adsorption. It is relatively easy to generate the higher carrier densities in the wider nanoribbons because of more available adsorption positions. The alkali-induced n-type doping have been observed in carbon-related systems, such as, graphite [89, 50], fullerene [115, 322], carbon nanotube [182, 284], and graphene [65, 258]. The electrical conductance is expected to be greatly enhanced in the increase of alkali concentration. The

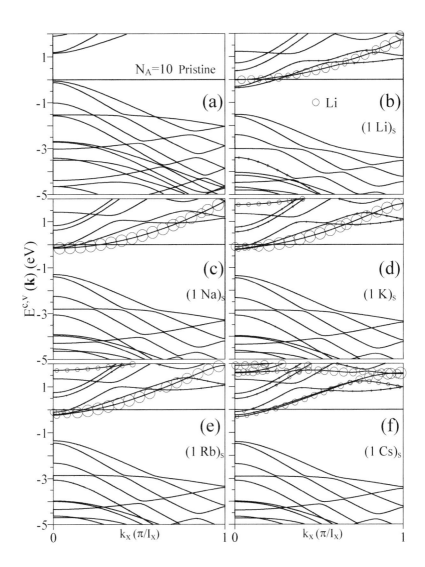

**FIGURE 9.2**
Band structures of $N_A = 10$ armchair graphene nanoribbons with hydrogen passivation for (a) pristine system and (b) Li-, (c) Na-, (d) K-, (e) Rb-, and (f) Cs-adsorbed ones. Only one adatom is situated at $(1)_s$.

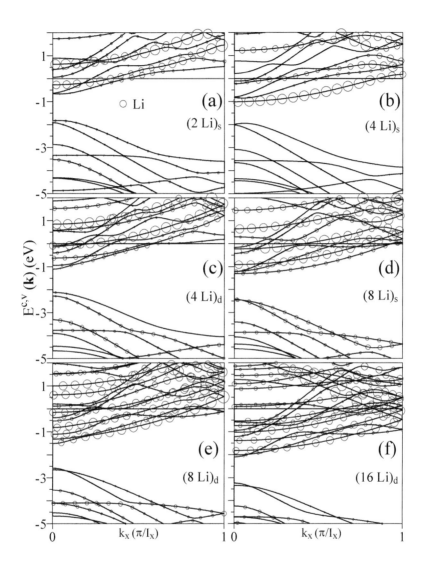

**FIGURE 9.3**
Energy bands of Li-adsorbed $N_A = 10$ armchair graphene nanoribbons under
various distributions and concentrations: (a) $(2\ \text{Li})_s$, (b) $(4\ \text{Li})_s$, (c) $(4\ \text{Li})_s$,
(d) $(8\ \text{Li})_s$, (e) $(8\ \text{Li})_d$, and (f) $(16\ \text{Li})_d$.

alkali-adsorbed graphene nanoribbons might have high potentials in nano-electronic devices [230].

The spatial charge distributions, which create the rich electronic properties, present the orbital hybridizations in chemical bonds. They are characterized by charge density ($\rho$) and the variation of charge density after adsorption ($\Delta\rho$), in which the latter is the density difference between adatom-adsorbed and pristine systems. A pristine graphene nanoribbon has the $\pi$ and the $\sigma$ bondings arising from the parallel $2p_z$ orbitals and the planar ($2s; 2p_x; 2p_y$) orbitals, respectively, as clearly indicated by the dashed and the solid rectangles in Figure 9.4(a). The m orbital hybridizations, with rather high charge density between two carbon atoms, belong to very strong covalent bonds in all the alkali-adsorbed systems (Figures 9.4(b)–9.4(f)). They remain almost the same under the alkali adsorptions, corresponding to slight variation of charge density ($\Delta\rho$ in Figures 9.4(g)–9.4(k)). However, the modified $\pi$ bondings could survive in a planar graphene nanoribbon for various configurations, concentrations and edge structures. The changes on the $\pi$ bondings are observable through the charge variations between carbon and alkali atoms (the heavy dashed rectangles), and they are relatively easily identified at high concentrations (Figures 9.4(i)–9.4(k)). These clearly indicate the single-orbital $s - 2p_z$ hybridization in the X-C bond. They could account for the absence of the alkali-dominated valence bands (Figures 9.2 and 9.3). At sufficient high concentrations, there exist obvious charge variations on x-y plane between two neighboring alkali atoms (the solid rectangles in Figures 9.4(i) and 9.4(k)), illustrating the significant $s - s$ orbital hybridization in A-A bonds.

The special structures in the orbital- and atom-decomposed DOSs are very useful in the further identifications of the critical orbital hybridizations in X-C and X-X bonds. Alkali adatoms can induce obvious changes in DOSs, especially for those at $E \geq 0$. An energy gap in pristine nanoribbon, being characterized by one pair of opposite-side anti-symmetric peaks (Figure 9.5(a)), become an energy spacing below the Fermi level $\sim 1$ eV after alkali adsorption (the blue-shift $E_F$). There are minor modifications on the prominent valence-state peaks due to the $2p_z$ (red curves) and $2p_x + 2p_y$ (blue curves) orbitals, as shown in Figures 9.5(b)–9.5(f). Furthermore, such asymmetric peaks are almost independent of alkali adatoms. However, the opposite is true for the strong conduction-state peaks. Their number, energy and intensity are very sensitive to alkali concentration and distribution. Some of asymmetric peaks above $E \geq -1$ eV are dominated by/related to alkali adatoms (green curves). They are merged with those of the $2p_z$ orbital (red curves), clearly illustrating the $s$-$2p_z$ hybridizations in X-C bonds. The alkali-dependent prominent peaks appear in the overall conduction-state range, when adatom concentrations are sufficiently highly, e.g., Figures 9.5(d)–9.5(f). A very wide continuous distribution directly reveals the $s$-$s$ orbital hybridization in X-X bonds. The alkali-induced structure variations in DOS and charge density differences could provide the consistent pictures in orbital hybridizations.

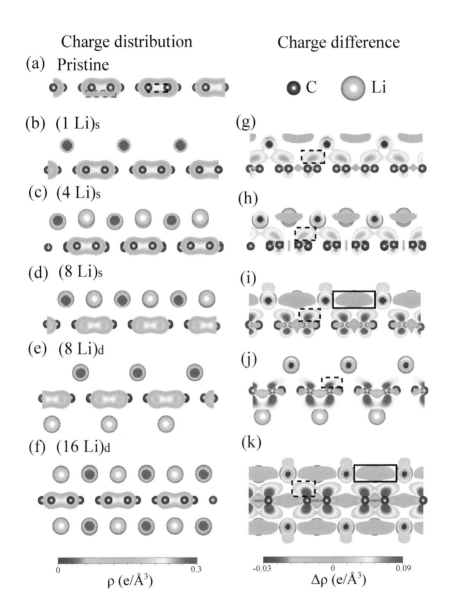

**FIGURE 9.4**

The spatial charge densities of Li-adsorbed $N_A = 10$ armchair nanoribbons under various adatom configurations: (a) pristine, (b) $(1,Li)_s$, (c) $(4,Li)_s$, (d) $(8,Li)_s$, (e) $(8,Li)_d$. The charge density differences, corresponding to (b)–(e), are shown in (g)–(j), respectively. (f) and (k) are those of $N_z = 10$ zigzag system under single adatom adsorption.

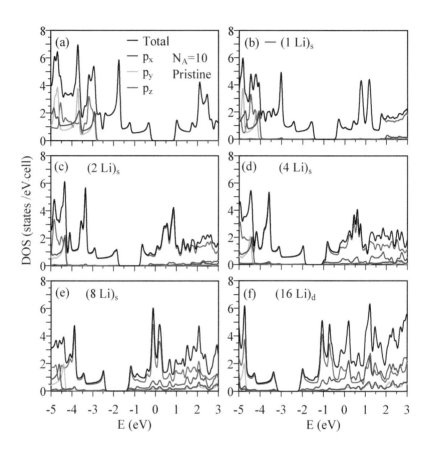

**FIGURE 9.5**
The atom- and orbital-projected DOSs of Li-adsorbed $N_A = 10$ armchair systems under various adatom configurations: (a) pristine, (b) $(1 \text{ Li})_s$, (c) $(2 \text{ Li})_s$, (d) $(4 \text{ Li})_s$, (e) $(8 \text{ Li})_s$, (f) $(16 \text{ Li})_d$.

## 9.2 The edge- and adsorption-co-dominated magnetic configurations

The edge-dependent band structures and magnetisms in zigzag graphene nanoribbons could be greatly diversified by alkali surface adsorptions. The low-lying energy bands exhibit the dramatic changes even for single-adatom adsorptions, as shown in Figures 9.6(a)–9.6(e). When one Li adatom is situated at the center of edge hexagon [$(1)_s$ in Figure 9.6(b)], the main features of the edge-localized partially flat bands are thoroughly changed, covering the splitting of spin-up and spin-down electronic states, and the intersecting with the Fermi level. The ↑- and ↓-dominated energy bands clearly indicate that the spin distribution in pristine zigzag nanoribbon (discussed later in Figure 9.7(a)), the anti-ferromagnetic ordering across the nanoribbon and the ferromagnetic configuration at each zigzag edge, is dramatically transformed by the significant orbital hybridization in X-C bond. The optimal spin configuration is determined by the competitions between the alkali-induced chemical bondings and the spin-dependent on-site Coulomb interactions [218]. The occupied free electron densities are different in the spin-split conduction bands (blue triangles), corresponding to the ferromagnetic spin configuration (Figure 9.7(b)). Another pair, which belongs to valence bands, are fully occupied, irrespective of magnetism. The spin splitting declines rapidly as Li gradually approaches to the nanoribbon center [$(3)_s$ and $(5)_s$ in Figures 9.6(b) and 9.6(c)]. A 1D metallic zigzag nanoribbon, with spin degeneracy, appears under the central adsorption [$(9)_s$ in Figure 9.6(d)]. Its spin configuration is similar to that of a pristine system (Figure 9.7(e)). The energy spacings/dispersions of the partially flat conduction and valence bands are affected by alkali adsorption; furthermore, part of electronic states in the former becomes occupied. Specially, these two edge-localized bands are merged together and possess fourfold degeneracy in the range of $k_x \leq 0.5$ for two adatoms at distinct edges, e.g., $(1,18)_s$ in Figure 9.7(f). The energy spacings, being created by the edge spin configurations, vanish mean the absence of magnetism. The destruction of spin distributions could be driven by the edge surface adsorptions. It should be noticed that the dramatic changes in the low-lying conduction bands do not alter the summation of the Fermi momenta/the total free carrier density under the same alkali concentration. By the detailed calculations and examinations, the alkali-induced conduction electron density is deduced to be independent of edge structure and magnetic configuration (Table 9.1(b)).

The spatial spin densities can present the diverse magnetic configurations in alkali-doped zigzag graphene nanoribbons. For a pristine system, the spin-up and spin-down configurations, respectively, appear in the A and B sublattices, as illustrated in Figure 9.7(a) by the red and blue balls. Their densities decrease quickly from zigzag edge to ribbon center, in which the former/the latter dominates the upper/lower edge spin distribution. They, respectively,

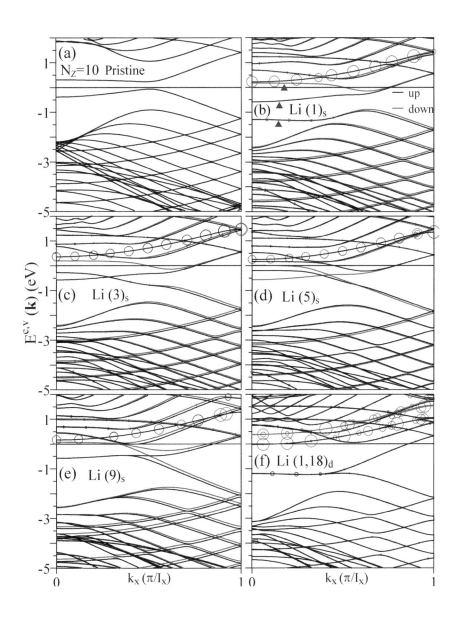

**FIGURE 9.6**
Band structures of $N_z = 10$ Li-adsorbed zigzag graphene nanoribbons for various adatom configurations: (a) pristine, (b) $(1)_s$, (c) $(3)_s$, (d) $(5)_s$, (e) $(9)_s$, and (f) $(1,18)_s$.

create the anti-ferromagnetic and ferromagnetic configurations along the longitudinal and transverse directions. As a result, there exist the separated edge-localized conduction and valence bands with spin-degenerate states (Figure 9.6(a)). The lower-edge spin distribution is fully destroyed by one alkali adatom there [$(1)_s$ in Figure 9.7(b)], so that only the spin-up-dominated distribution survives in the upper edge. This is responsible for the spin-split free electrons in conduction bands (Figure 9.6(b)). The spin-down-dependent distribution gradually recovers as adatom leaves from edge [$(3)_s$ and $(5)_s$ in Figures 9.7(c) and 9.7(d)]. The central adsorption is similar to the pristine case in terms of magnetic configuration [$(9)_s$ in Figure 9.7(e)]. However, the $s$-$2p_z$ hybridization in X-C bond changes the semiconducting into the metallic behavior (Figure 9.6(e)). Specially, spin configuration becomes absent for two adatoms at the upper and lower edges (not shown). The non-magnetic 1D metals are expected to be relatively easily observed in the high-concentration alkali-adsorbed zigzag graphene nanoribbons.

The alkali-enriched magnetic configurations are directly reflected in the low-energy DOS structures, as clearly indicated in Figures 9.8(a)–9.8(f). These are the main differences between the zigzag and armchair systems. The former can present the delta-function-like symmetric peaks due to the edge-localized energy bands, strongly depending on the spin configurations. The ferromagnetic, anti-ferromagnetic and non-magnetic alkali-adsorbed zigzag graphene nanoribbons, respectively, exhibit three peaks (blue triangles in Figures 9.8(b)–9.8(d)), a pair of symmetric peaks below the Fermi level (Figure 9.8(e)), and a merged peak (Figure 9.8(f)). The peak intensities are, respectively, and enhanced in the spin-split and merged energy bands (Figures 9.6(b)–9.6(d) and Figure 9.6(f)). It might have some weaker peaks between them (red circles), since the edge-localized energy bands possess an extra band-edge state at larger $k_x$. Apparently, the number, energy and intensity of the low-energy symmetric peaks are mainly determined by the magnetic configurations. The STS measurements on the main features of DOS could be utilized to identify the zigzag-edge spin distributions under the alkali surface adsorptions.

Up to now, there are no experimental examinations for the unusual energy bands and the diverse magnetic configurations in alkali-adsorbed graphene nanoribbons. The ARPES/spin-resolved ARPES measurements are required to verify the occupied valence and conduction bands below the Fermi level, including the smallest energy spacing between them, the alkali-dominated conduction bands, the carbon-related conduction and valence bands, and the spin-split/spin-degenerate edge-localized energy bands in zigzag nanoribbons. They are available in identifying the $\pi$ and $\sigma$ bondings in C-C bond, the $s$-$2p_z$ hybridization in X-C bond, and the $s$-$s$ hybridization in X-X bond. Specifically, the experimental verifications on the finite/vanishing energy spacing and spin splitting/degeneracy in zigzag systems are associated with the ferromagnetic/anti-ferromagnetic/non-magnetic configurations. The critical orbital hybridizations and the diverse spin distributions could also be identi-

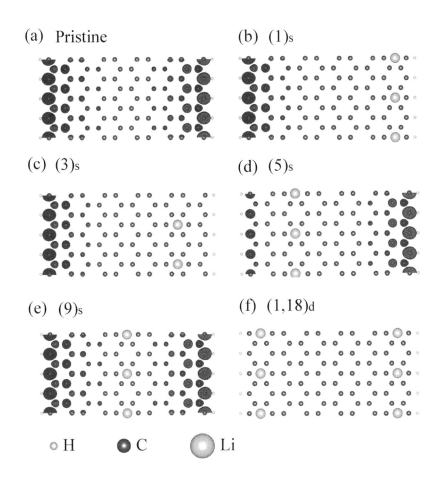

**FIGURE 9.7**
The spatial spin distributions for $N_z = 10$ Li-adsorbed zigzag graphene nanoribbons under various adatom configurations: (a) pristine, (b) $(1)_s$, (c) $(3)_s$, (d) $(5)_s$, (e) $(9)_s$, and (f) $(1,18)_s$

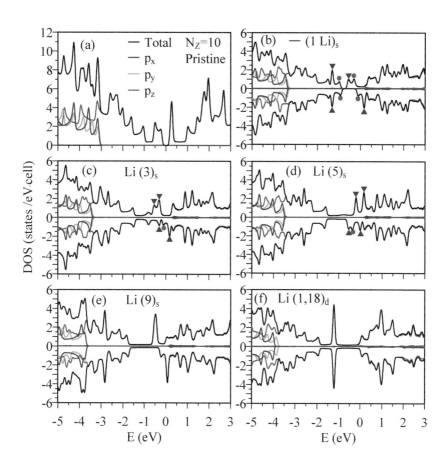

**FIGURE 9.8**

The atom- and orbital-projected DOSs of Li-adsorbed $N_z = 10$ zigzag systems for various adatom configurations: (a) pristine, (b) $(1)_s$, (c) $(3)_s$, (d) $(5)_s$, (e) $(9)_s$, and $(1,18)_s$.

fied from the STS/spin-polarized STS measurements on the special structures of DOS, covering the main features of asymmetric peaks and delta-function-like symmetric peaks in the overall energy range. Among three kinds of magnetic configurations, the ferromagnetic energy bands can create the spin-split absorption spectra in magneto-optical Kerr measurements [94, 67] and the spin-polarized currents in transport measurements. The latter might have potential applications in spintronic devices [118, 99].

# 10

## Halogen-adsorbed GNRs

Halogens are quite different from alkalis in the valence electrons of the outermost orbitals, and so do the adsorption effects on the essential properties. The halogenated graphenes are verified/predicted to exhibit the rich and unique geometric, electronic and magnetic properties, especially for the fluorinated systems. The full fluorination of graphene can destroy the Dirac-cone band structure and create a direct energy larger than 3 eV [252, 240]. Furthermore, the metallic behavior might appear at certain low-concentration distributions [252]. These clearly show the strong competitions between the $\pi$ and F-C chemical bondings. The tunable electronic properties have been identified to display the high potentials in various applications, such as, lithium batteries [328], supercapacitors [393], molecule detectors [154], and biosensors [343]. However, the (Cl,Br,I,At)-adsorbed graphenes must belong to the p-type metals with the slightly modified conducting $\pi$ bondings [252, 74]. In addition to the NM configuration, the FM and AFM spin distributions are also revealed under specific distributions and concentrations [252]. All the halogen adatoms might induce the spin states after the chemical adsorptions. Based on the extra quantum-confinement and edge-structure effects, the halogenation effects of GNRs are expected to become more complicated and unique. Electronic structures and magnetic configurations are relatively easily diversified during the variations of adsorption cases; that is, the 1D halogenated systems will present more kinds of essential properties. Specifically, the relation between the linear hole density and adatom concentration could be exactly evaluated from the 1D energy bands.

The various halogenation effects on GNRs could be understood from five electrons in the outermost orbitals using the first-principles calculations. The critical orbital hybridizations in X-C, C-C, and X-X bonds and the diverse spin configurations due to halogen adatoms and zigzag-edge carbons are explored thoroughly. Specifically, the latter are identified from the magnetic moment, the spin-split energy bands associated with adatoms and/or carbons, the spatial spin distribution, and the spin-up- and spin-down-decomposed DOSs. The X-C bonding strength, which determines the relations between the chemical adsorptions and the electronic properties, is reflected in the bond length, the binding energy, the change of the $\pi$- and $\sigma$-electronic band structures, the (X,C)-co-dominated energy bands, the variation of the spatial charge density, and the merged prominent peaks in DOSs. The X-C bonds and the X-induced spin states will be utilized to distinguish the fluorination and chlorination-related effects, such as, the semiconducting or metallic behavior, the existence of a simple relation between the free hole density and adatom concentration, and the dependence of FM/AFM/FM configurations on concentrations, distributions. and edge structures. How many kinds of electronic and magnetic properties are, respectively, revealed in fluorinated and other halogenated systems is the main focus.

## 10.1   Fluorination effects

The fluorination effects induce the unusual geometric properties clearly illustrated in Table 10.1. The top site is the optimal F-adatom position, regardless of adsorption cases, edge structures and ribbon widths (Figure 10.1). In general, the double-side/higher-concentration adsorptions have the larger binding energies, compared with the single-side/lower-concentration cases. With F adsorptions, the passivated carbon atoms deviate from the GNR plane, being sensitive to distributions and concentrations. Their heights vary from 0.320 Å to 0.542 Å, depending on the single-/double-side adsorption and the adatom concentration. Apparently, the fluorinated GNRs exhibit the buckled honeycomb lattices. F adatoms are very close to carbon atoms, in which the shortest and longest F-C bond lengths, 1.384 Å and 1.563 Å, respectively, correspond to the highest and lowest concentrations. Moreover, the nearest C-C bonds are lengthened in the range of $0.056-0.135$ Å, as measured from those of pristine systems. These indicate that the strong F-C bondings are responsible for the featured geometric structures and are deduced to modify the $\pi$ and $\sigma$ bondings of carbon atoms. The F-C bonds will play critical roles in other essential properties. In short, the fluorine- and alkali-adsorbed systems sharply contrast with each other in the binding energy, the planar/non-planar structure, the optimal position; the X-C and C-C bond lengths.

The fluorine adsorptions dramatically change the main features of electronic structures (Figure 10.2), including band gap, free carrier density, energy dispersion, subband spacing band-edge states, atom dominance, and spin degeneracy. Concerning the single-adatom case, all the F-adsorbed armchair GNRs exhibit the metallic band structures, as shown in Figures 10.2(b) and 10.2(c) (also in Table 10.1). The Fermi level is shifted to the C-dominated valence states (Figures 10.2(a) and 10.2(b)), i.e., $E_F$ exhibits a red shift, or the single F adsorption can creates the p-type doping. There exist free holes in the unoccupied  valence states between two Fermi momenta, since electrons are transferred from carbon atoms to fluorine adatoms. That fluorine possesses a very strong electron affinity is the main reason. The low-lying energy bands mainly originate from the conducting $\pi$ bondings of carbon atoms, being almost independent of adatom positions. Their energy dispersions and spacings are drastically changed after fluorination. Moreover, the (F,C)-co-dominated

**TABLE 10.1**
Binding energy, magnetic moment/magnetism, metal/energy gap, free hole density, X-C bond length, deviation height of C and the first C-C bond length for halogenated $N_A = 10$ armchair and $N_Z = 10$ zigzag GNRs under various distributions and concentrations.

| GNRs | Systems | $E_b$ (eV) | $M_{tot}$ ($\mu B$) | $E_g$ (eV) | # of holes | X-C (Å) | C dev. (Å) | Nearest C-C (Å) |
|------|---------|-----------|---------------------|-----------|-----------|---------|-----------|------------------|
| AGNR | H-terminated | - | 0/NM | 1.112 | 0 | - | 0 | 1.428 |
| | Cl-$(1)_s$ | -0.8232 | 0.74/FM | M | 1 | 3.345 | 0 | 1.430 |
| | Cl-$(6)_s$ | -0.8265 | 0.65/FM | M | 1 | 3.345 | 0 | 1.430 |
| | Cl-$(6,13)_s$ | -0.7932 | 1.28/FM | M | 1 | 3.357 | 0 | 1.433 |
| | Cl-$(6,13)_d$ | -0.8166 | 1.28/FM | M | 1 | 3.357 | 0 | 1.433 |
| | Cl-$(4F)_d$ | -0.9140 | 2.46/FM | M | 2 | 3.362 | 0 | 1.450 |
| | Cl-$(6F)_d$ | -0.9521 | 5.12/FM | M | 3 | 3.368 | 0 | 1.458 |
| | Cl-$(8F)_d$ | -1.0289 | 5.01/FM | M | 4 | 3.372 | 0 | 1.475 |
| | Cl-$(10F)_d$ | -0.9812 | 2.39/FM | M | 5 | 3.378 | 0 | 1.483 |
| | Cl-$(12F)_d$ | -0.9932 | 0/NM | M | 6 | 3.383 | 0 | 1.512 |
| | Cl-$(14F)_d$ | -0.9528 | 0/NM | M | 7 | 3.388 | 0 | 1.532 |
| | Cl-$(16F)_d$ | -0.9627 | 0/NM | M | 8 | 3.392 | 0 | 1.557 |
| | Cl-$(20F)_d$ | -0.9476 | 0/NM | M | 10 | 3.457 | 0 | 1.573 |
| | Cl-$(20F)_d$ | -0.6530 | 0/NM | 2.244 | 0 | 1.784 | 0.327 | 1.732 |
| | Br-(6) | -0.5496 | 0.69/FM | M | 1 | 3.615 | 0 | 1.429 |
| | Br-$(6,13)_d$ | -0.5021 | 1.41/FM | M | 1 | 3.684 | 0 | 1.430 |
| | I-(6) | -0.4585 | 0.52/FM | M | 1 | 3.750 | 0 | 1.430 |
| | At-(6) | -0.3398 | 0.19/FM | M | 1 | 4.107 | 0 | 1.430 |
| ZGNR | H-terminated | - | 0/AFM | 0.374 | 0 | - | 0 | 1.428 |
| | Cl-$(3)_s$ | -1.2795 | 0.29/FM | M | 1 | 3.009 | 0 | 1.432 |
| | Cl-$(19)_s$ | -1.3196 | 0.51/FM | M | 1 | 3.009 | 0 | 1.430 |
| | Cl-$(3,19)_d$ | -0.8631 | 0.67/FM | M | 1 | 3.002 | 0 | 1.430 |
| | Cl-$(19,22)_d$ | -0.8138 | 0.14/FM | M | 1 | 3.002 | 0 | 1.430 |
| | Cl-$(3,38)_d$ | -1.0535 | 0/AFM | M | 1 | 3.002 | 0 | 1.430 |
| | Br-$(19)_s$ | -0.8219 | 0.92/FM | M | 1 | 3.600 | 0 | 1.430 |
| | I-$(19)_s$ | -0.5981 | 1.04/FM | M | 1 | 3.823 | 0 | 1.430 |
| | At-$(19)_s$ | -0.3822 | 1.12/FM | M | 1 | 3.900 | 0 | 1.430 |

Note: NM, FM and AFM correspond to non-magnetism, ferro-magnetism and anti-ferro-magnetism, respectively.

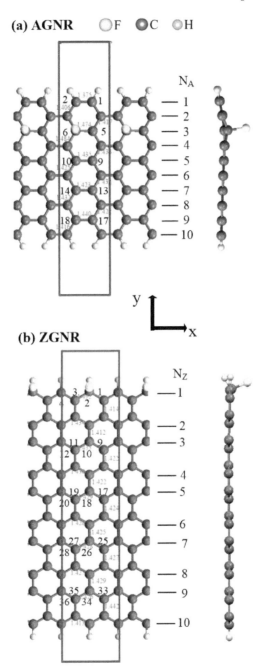

**FIGURE 10.1**
Geometric structures of F-adsorbed GNRs for (a) $N_A = 10$ armchair structure
and (b) $N_Z = 10$ zigzag one. The red rectangles represent unit cells. The yellow
color represents the F atoms adsorbed on the top-side of GNRs.

energy bands, accompanied with the modified $\sigma$ bands, come to exist at deeper energy of $E^v \leq -2$ eV. Such bands have the weak energy dispersions and the narrow band widths. The feature-rich energy bands clearly illustrate that the critical F-C bondings induce the significant modifications on $\pi$ and $\sigma$ bands ($\pi$ and $\sigma$ bondings). Specifically, the spin-split band structures might appear under certain distributions, e.g., the edge $(1)_s$ adsorption in Figure 10.2(c). The low-lying energy bands across the Fermi level determine the magnetic moment/FM configuration (Table 10.1 and Figure 10.4(a)), in which they are dominated by carbon atoms, but not F adatoms. The F-adatom adsorption is deduced to induce the spin distribution of the neighboring carbons, since the occupied spin-up and spin-down states strongly depend on the C-dominated low-lying energy bands. This will diversify the spatial spin distributions or create the diverse magnetic configurations.

Band structures are greatly enriched by the concentration, relative position, and edge structure, but not single- or double-side adsorptions. With the gradual increment of concentration, the two and four-F systems might belong to the 1D p-type metals, as shown in Figures 10.2(d)–10.2(f). The total free hole density, the summation of the Fermi momenta in partially unoccupied $\pi$-electronic valence bands, becomes higher $[(6,13)_s$ in Figure 10.2(d), $(6,13)_d$ in Figure 10.2(e) and $(1,6,13,20)$ in Figure 10.2(h); Table 10.1] or keeps the same $[(1,19)_d$ in Figure 10.2(f)], compared with that of the single-F case (Figures 10.2(b) and 10.2(c)). On the other hand, the metallic band structure might be thoroughly transformed into the semiconducting one for two close adatoms, e.g., an indirect gap of 0.96 eV under the $(1,6)_d$ adsorption (Figure 10.2(g)). This suggests the termination/serious distortion of the extended $\pi$ bonding in the honeycomb lattice. The semiconducting behavior appears more frequently in the further increase of F-concentration, mainly owing to the great enhancement of the F-C bonding effects $((4F)_d$-case in Figure 10.2(i) and $(8F)_d$-case in Figure 10.2(j)). The critical concentration in which all the F-doped AGNRs belong to the semi-conducting systems is estimated to be over 50% (F:C=10:20). Specifically, the high-concentration semiconductors do not present the spin-split energy bands (Figures 10.2(j)–10.2(l)) and belong to the NM systems (Table 10.1). This reveals the magnetic suppression closely related to the strong F-F bonds (discussed later). These systems might have large direct/indirect energy gaps, e.g., $E_g^d = 3.4$ eV under the full fluorination. Such gaps are mainly determined by the $\sigma$-orbital hybridizations of C-C bonds, and the multi-orbital hybridizations in F-C and F-F bonds, since the conducting $\pi$ bonding is absent. In addition, the single- and double-side adsorptions present the almost identical low-lying energy bands (e.g., Figures 10.2(d) and 10.2(e)) under the specific (x,y) projection and concentration.

The strong competition/cooperation between fluorination and zigzag edge further diversifies the essential properties. There exist drastic changes on electronic structures, especially for the spin degeneracy, number, energy gap and hole density of the partially flat bands (Figures 10.3(a)–10.3(f)). When one F adatom is situated at the zigzag edge, the number of edge-localized energy

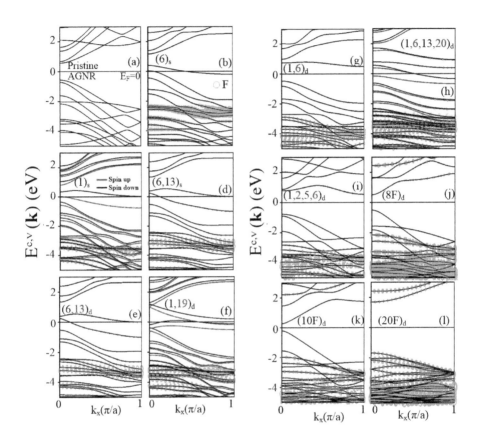

**FIGURE 10.2**
Band structures of $N_A = 10$ AGNRs for (a) H-terminated, (b) $(6)_s$-, (c) $(1)_s$-, (d) $(6,13)_s$-, (e) $(6,13)_d$-, (f) $(1,19)_d$-, (g) $(1,6)_d$-, (h) $(1,6,13,20)_d$-, (i) $(1,2,5,6)_d$-, (j) $(8F)_d$-, (k) $(10F)_d$-, and (l) $(20F)_d$-adatom adsorptions. The green circles represent the contribution of F adatoms. The black and red curves denote the spin-split energy bands. The subscripts s and d, respectively, correspond to the single- and double-side adsorptions, while the black and red numbers imply the top and bottom adsorption positions.

bands becomes half in the presence of spin splitting, as revealed under the $(3)_s$ adsorption in Figure 10.3(b). Such bands intersect with the Fermi level and possess free holes (Table 10.1), in which the number of occupied states is different in the spin-up and spin-down components. This system, with a net magnetic moment $(0.768\ \mu_B)$, belongs to a 1D FM metal. Specifically, the partially flat bands disappear under two adatoms near distinct zigzag edges [e.g., $(3,38)_d$ in Figure 10.3(d)]; furthermore, a direct energy gap appears at zone boundary in the absence of spin splitting and magnetism (a NM semiconductor with $E_g^d = 0.27$ eV). The strong F-C bands could destroy localization behavior and spin configuration due to the edge carbons. But for the central two-F adsorption, the partially flat bands, with spin splitting, become double $[(11,14)_d$ in Figure 10.3(e)], in which half of them correspond to the edge- or center-localized electronic states. An indirect narrow gap $(E_g^i = 0.17$ eV) is accompanied with the vanishing magnetic moment, indicating a AFM semiconductor. On the other side, the number of low-lying flat bands remain unchanged, when two adatoms are, respectively, close to edge and center $[(3,19)_d$ in Figure 10.3(f)]. The spin splitting means a 1D FM metal.

The fluorinated GNRs possess four kinds of spin-dependent electronic and magnetic properties. The spin-split energy bands strongly depend on fluorine adsorption and edge structure. The H-terminated armchair systems belong to the confinement-induced semiconductors without spin splitting in band structure and magnetism. The NM semiconductors are frequently revealed in fluorinated systems especially for the higher-concentration cases (the first kind in Figures 10.2(i)–10.2(l) and Table 10.1). As to the metallic systems, they present the spin-degenerate energy bands in the absence of magnetism (the second kind in Figures 10.2(b), 10.2(d), 10.2(e), and 10.3(d)), or the spin splitting under the FM configuration (the third kind in Figures 10.2(c), 10.2(f), 10.2(h), 10.3(b), 10.3(c), and 10.3(f); discussed later in Figure 10.4). In addition to NM semiconductors and FM metals, the fluorinated zigzag systems might be the AFM semiconductors with spin splitting (the fourth kind in Figures 10.3(e) and Figure 10.4(f)). However, the spin splitting is absent in the AFM H-passivated zigzag systems (the fifth kind in Figure 10.3(a)).The ARPES and spin-resolved ARPES are available in examining the predicted five kinds of electronic structures and magnetic configurations. That is to say, the complicated relations among the finite-size effects, the edge structures, the spin configurations, and the multi-orbital hybridizations could be directly examined from such measurements.

The p-type doing in NM and FM metals deserves a closer investigation. A simple relation between the fluorination-induced hole density and adatom concentration is absent except for one-F case, according to the delicate calculations and detailed analyses. A single F in a unit cell can create one free hole for distinct adsorption positions (Table 10.1). One electron is transferred from the bonded carbon atom to the adatom under a rather strong fluorine affinities. AGNRs could create one and two holes per unit cell below 50%, respectively, corresponding to the spin-split and spin-degenerate p-electronic

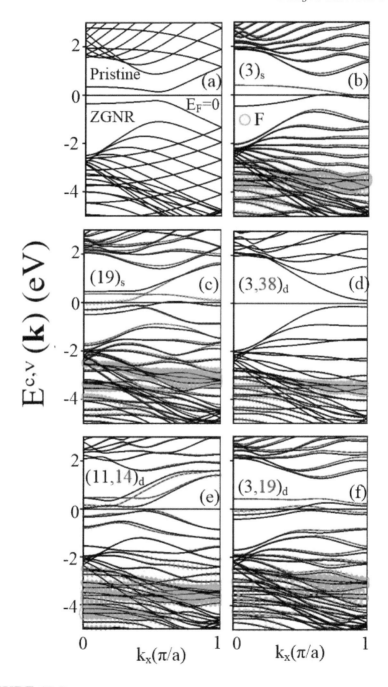

**FIGURE 10.3**
Band structures of $N_Z = 10$ ZGNRs for (a) H-terminated, (b) $(3)_s$, (c) $(19)_s$, (d) $(3,38)_d$, (e) $(11,14)_d$, and (f) $(3,19)_d$ cases.

energy bands in Figure 10.2 (FM and NM configurations in Table 10.1). However, ZGNRs only exhibit the lower carrier-density case (Figures 10.3(b) and
10.3(f); Table 10.1). The fluorination and alkalization effects quite differ from
each other in the metallic behavior. The latter belongs to the n-type doping
even for the the full adsorption, regardless of adatom concentration. Each alkali adatom generates one conduction electron from the outermost s-orbital
by means of the significant alkali-C bond, regardless of adatom distributions.
This will result in very high conduction electron density. However, the strong
F-C bonds could destroy/seriously distort the conducting $\pi$ bonding, leading to the great reduce of hole density and even creating the semiconducting
behavior.

The spatial spin densities and magnetic moments could provide more information about magnetic properties. They can characterize the FM configurations and the AFM ones with and without the spin-split energy bands,
respectively. The spin arrangements are sensitive to the adatom distribution
and edge structure, as revealed in Figures 10.4(a)–10.4(f). Furthermore, the
competition between the spin-up and spin-down components determines the
net magnetic moment in a unit cell (Table 10.1). The spin densities are mostly
distributed near the edge structures. When one F is situated at the armchair
edge, it creates the spin-up and spin-down distributions of the neighboring carbons (the red and blue colors in Figure 10.4(a)). The former dominates the FM
configuration and their competition leads to a net magnetic moment of 0.61
$\mu_B$ ((1)s in Table 10.1). The FM is greatly enhanced for two adatoms near distinct boundaries, e.g., $(1,19)_d$ and $(1,6,13,20)_d$ ($0.87\mu_B$ in Figure 10.4(b) and
1 $\mu_B$ in Figure 10.4(c)). As to the zigzag structure, the edge-F adatom could
suppress the intrinsic carbon-spin states, but induce the dominating spin-down
configuration at another edge. For example, the $(3)_s$ adsorption presents the
FM configuration with magnetic moment 0.77 $\mu_B$ (Figure 10.4(d)). There is
no magnetism for two adatoms at the upper and lower zigzag edges, e.g., the
$(3,38)_d$ adsorption (Table 10.1). Specifically, the AFM configuration, without
the central distribution, in a H-terminated system (Figure 10.4(e)) becomes
more complex under the central two-F adsorption (Figure 10.4(f)). The coex-
istence of the edge- and center-initiated spin distributions indicates an unusual
AFM configuration without magnetic moment.

The charge density, charge density difference, and partial charge density
can directly identify the multi-orbital hybridizations in F-C, C-C, and F-F
bonds and the semiconducting/metallic behavior. There exist much charge
density between fluorine and carbon, as indicated from H-terminated and
fluorinated systems (Figures 10.5(a), 10.5(b), and 10.5(d)). Furthermore, fluorination leads to an obvious change in the $\pi$ bonding and an observable
decline in the $\sigma$ bonding. The strong fluorination effects also induce the drastic density variations, especially for $\Delta\rho$ near F adatoms on $(x, z)$ and $(y, z)$
planes (Figures 10.5(c) and 10.5(e)). These clearly indicate the complicated
$(2p_x, 2p_y, 2p_z) - (2p_x, 2p_y, 2p_z)$ hybridizations in F-C bonds. As to F-F bonds,
the significant $(2p_x, 2p_y)$ hybridizations appear for two neighboring fluorine

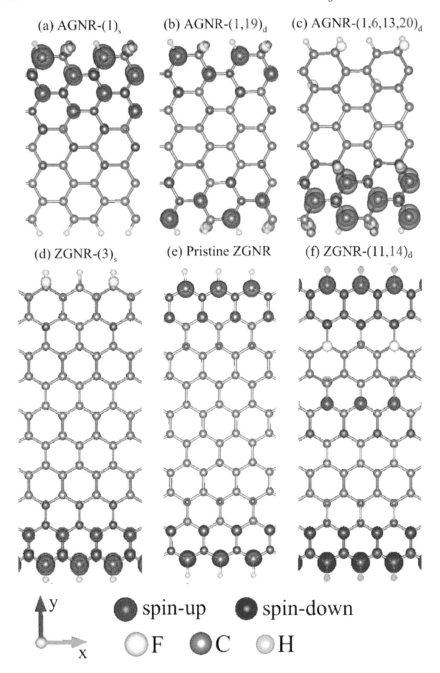

**FIGURE 10.4**
Spin density distributions of $N_A = 10$ fluorinated AGNRs under (a) $(1)_s$, (b) $(1,19)_d$, (c) $(1,6,13,20)_d$ cases, and $N_Z = 10$ ZGNRs for (d) $(3)_s$, (e) H-terminated, and (f) $(11,14)_d$ cases.

adatoms, as illustrated in Figures 10.5(c) and 10.5(e) (red rectangles on $(y,z)$ planes). The partial charge density, which corresponds to electronic states very close to the Fermi level, might exhibit the continuously extended and the discontinuously terminated $\pi$ bondings. The former and the latter, which are, respectively, revealed in Figures 10.5(f) and 10.5(g) and Figures 10.5(h) and 10.5(i), are responsible for the metallic and semiconducting properties. That is, whether the $\pi$ bonding could continuously extend in the GNR plane determines this property.

The diverse band structures with/without spin splitting are directly reflected in the orbital- and spin-projected DOSs (Figure 10.6), further illustrating the orbital hybridizations and magnetic configurations. A lot of asymmetric and symmetric peaks are, respectively, associated the parabolic and partially flat bands. When the F-concentration is below 50 % (Figures 10.6(a)–10.6(c) and 10.6(e)–10.6(j)), the low-energy DOS is dominated by the distorted $\pi$ bonding of C-$2p_z$ orbitals (red curves). Such bonding also makes significant contributions to the deeper-energy DOS. The peak structures due to the $\sigma$ bonding of C-$(2p_x,2p_y)$ orbitals appear at $E < -2$ eV (blue curves). The similar structures are induced by the $(2p_x,2p_y)$ and $2p_z$ orbitals of the F adatoms (pink and light blue curves). The former present a sufficiently wide energy width of $\sim 2$ eV; that is, there exist the significant $(2p_x,2p_y)$ orbital hybridizations in F-F bonds. Specially, there are some differences for the peak structures dominated by the C-$2p_x$ and C-$2p_y$ orbitals, mainly owing to the 1D quantum confinement effects (insets in Figures 10.6(a) and 10.6(f)). All the orbitals contribute to the merged peak structures at deeper energy, being attributed to the $(2p_x,2p_y,2p_z)$-$(2p_x,2p_y,2p_z)$ multi-orbital hybridizations in F-C bonds. As for high-concentration DOSs, the energy widths of the F-induced bands might be more than 3 eV (Figures 10.6(c)–10.6(d)); furthermore, the corresponding peaks are stronger than those of the $\sigma$ bands. The occupied valence states in the range of $E \geq -5$ eV mainly arise from the $(2p_x,2p_y)$ orbitals of F and C. This indicates the serious destruction of the $\pi$ bonding.

Five kinds of electronic and magnetic properties in fluorinated and H-passivated systems are characterized by the low-energy DOS peak structures. A pair of antisymmetric peaks across the Fermi level, being divergent in the opposite direction, is characteristic of energy gap in the absence of spin splitting (NM semiconductors; the first kind), as shown in Figures 10.6(a)–10.6(d). As to NM metals (the second kind), a similar pair presents a blue shift of $\sim 0.5 - 1.0$ eV above $E = 0$ (Figures 10.6(e) and 10.6(f)), in which DOS is finite at the Fermi level. The spin-polarized peak structures are revealed in FM metals (the third kind in Figures 10.6(g) and 10.6(h)). They are quite different for spin-up and spin-down configurations; furthermore, the former predominates the occupied states of $E < 0$. Specifically, the partially flat bands in fluorinated zigzag systems could create more pairs of symmetric peaks centered about $E = 0$ in the presence of spin splitting, as indicated in Figure 10.6(i) by purple circles. The fourth kind of peak structures originates from AFM narrow-gap semiconductors. A H-terminated zigzag system only induces

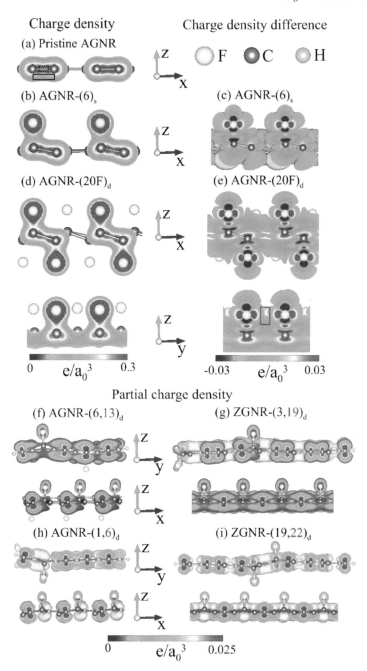

**FIGURE 10.5**
The spatial charge density of $N_A = 10$ AGNRs for (a) H-terminated, (b) $(6)_s$, and (d) $(20F)_d$ cases; the charge density difference for (c) $(6)_s$, and (e) $(20F)_d$ cases. The partial charge density is shown for (f) armchair-$(6,13)_d$, (g) zigzag-$(3,19)_d$, (h) armchair-$(1,6)_d$, and (i) zigzag-$(19,22)_d$.

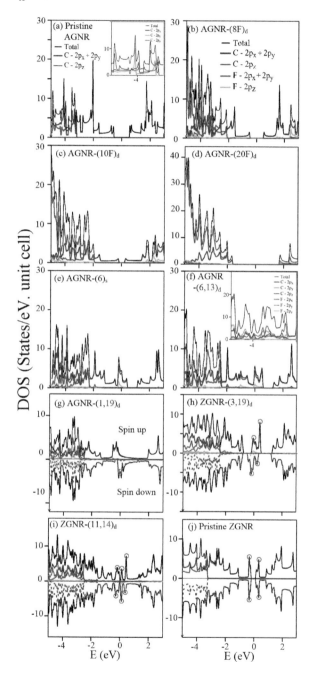

**FIGURE 10.6**

Orbital- and spin-projected DOSs for (a) pristine, (b) $(8F)_{d^-}$, (c) $(10F)_{d^-}$, (d) $(20F)_{d^-}$, (e) $(6)_{s^-}$, (f) $(6,13)_{d^-}$, (g) $(1,19)_{d}$-adsorbed $N_A = 10$ armchair systems; (h) $(3,19)_{d^-}$, (i) $(11,14)_{d}$-adsorbed and (j) H-terminated $N_Z = 10$ zigzag systems. Purple triangles arise from the partially flat bands.

a pair of symmetric peaks (Figure 10.6(j)), accompanied with an energy gap and one peak due to an extra band-edge state (Figure 10.3(a)). This represents an AFM semiconductor without spin degeneracy (the fifth kind). The above-mentioned five kinds of low-lying DOSs could be examined by the STS and SP-STS measurements on energy gap, a finite value at $E_F$, the spin-polarized and edge-localized peaks.

## 10.2    Chlorination-related systems

The Cl-, Br-, I-, and At-adsorbed GNRs quite differ from the fluorinated systems in essential properties. The former remain the planar structures (Table 10.1); furthermore, they are much higher than the latter ($\sim 3.5$ Å versus $\sim 1.5$ Å). These clearly indicate the weak but significant orbital hybridizations in non-fluorinated X-C bonds. The F-C bonds create the highest binding energy among all the systems. For example, $E_b$ changes from $\sim -3.54$ eV to $\sim -0.34$ eV for a single adatom in $N_A = 10$ armchair system as F→At. The neatest C-C bond length gradually grows in the increase of halogen concentration. It should be noticed that the Cl-, Br-, I- and At-absorbed graphenes possess the meta-stable configurations with the adatom heights close to those of F. This configuration presents a smaller binding energy compared to the most stable one, e.g., the chlorinated systems (Table 10.1). The STM and TEM measurements could be utilized to examine the predicted geometric properties, covering the planar/buckled structure, the height and position of halogen adatom, and the nearest C-C bond length. The single- or multi-orbital hybridization and coupling strength in chemical bonds will determine the diversified electronic properties and magnetic configurations.

The other halogen adatoms, which possess the similar effects on electronic properties, dramatically alter the low-lying energy bands. We first discuss the single-adatom adsorption. The original valence and conduction bands due to carbon honeycomb lattice almost keep the same after surface adsorptions, as revealed from a direct comparison between Figure 10.7(a) and Figures 10.7(b)–10.7(e). Specially, the Cl-related adatoms create four spin-split energy subbands (the green circled curves), in which most of them present very weak dispersion relations. Energy dispersions might become strong in the increment of atomic number. The Fermi level has red shift of $\sim 0.8$ eV, compared with the H-terminated system. Furthermore, it intersects with one carbon-dominated valence band and the spin-down energy bands (Cl, Br, and I in Figures 10.7(b), 10.7(c), and 10.7(d)) or the spin-split energy bands (At in Figure 10.7(e)). The crossing behaviors with $E_F$ depend on the kind of halogens, while the summation of the spin-up and spin-down Fermi momenta (the linear hole density) is almost identical for various adatoms. For example, the different hole densities in the unoccupied spin-up and spin-down states (red and black curves) represent the ferromagnetic spin configuration (discussed later in Figure 10.10). Apparently, the important differences between the Cl-related and fluorinated systems (Figures 10.7(b)–10.7(f)) lie in the drastic changes of the low-lying spin-split energy bands or the valence bands.

The halogenation effects on energy bands of armchair systems are very sensitive to adatom concentrations. For two adatom adsorption (Figures 10.8(a) and 10.8(b)), more valence states become unoccupied and change into free holes, including two carbon-dominated valence bands and two spin-down Cl-related energy bands. The total hole density is double that of single-adatom adsorption, and so does the FM (magnetic moment in Table 10.2). With the increase of Cl concentration (Figure 10.8(c)), there are more spin-split Cl-related energy bands, accompanied with a wider distribution range. Specifically, the spin splitting disappears under concentrations larger than 50% (Figures 10.8(d)–10.8(f)); that is, the net magnetic moment is vanishing above 50% without in Table 10.2). A lot of Cl-induced conduction and valence bands exist in a wide energy range. They are closely related to the great enhancement of the multi-orbital hybridizations in the Cl-Cl bonds (discuss later in Figure 10.10). Such strong bondings can fully suppress the ferromagnetic spin configurations.

The essential properties of ZGNR are greatly diversified by various concentrations and distributions of halogen adatoms. At low concentrations, the energy bands near $E_F$ are very sensitive to the adatom positions, as indicated in Figures 10.8(h)–10.8(i) and Figures 10.8(j)–10.8(l) for one- and two-adatom adsorptions, respectively. All the low-lying energy bands present the spin splitting, especially for the edge-carbon- (the red and black curves) and Cl-dominated bands with weak energy dispersions. This directly reflects the inequivalent magnetic environments of the spin-up and spin-down distributions (Figures 10.9(g), 10.9(i), and 10.9(j)). When the occupied state densities are not identical in the spin-up and spin-down energy bands, the FM configuration is created by the Cl-adsorption, as observed in Cl-$(3)_s$, Cl-$(19)_s$, Cl-$(3,19)_d$ and Cl-$(19,22)_d$ (Figures 10.8(h)–10.8(k)). The AFM configurations with equivalent spin-up and spin-down contributions are observed for the H-passivated ZGNR (Figure 10.8(g)) and two adatoms at distinct zigzag edges (Cl-$(3,38)_d$ in Figures 10.8(l)).

All the Cl-related halogenated systems belong to 1D $p$-type metals for different concentrations, distributions and edge structures, as clearly indicated in Table 10.2. For the concentration lower than 10%, one halogen adatom in a unit cell can provide one free hole. However, the free carrier density is reduced to half under the higher concentration. The Cl-related GNRs are predicted to create the highest-density free holes, compared with the other 1D condensed-matter systems. Such simple relations is absent in fluorinated systems, since most of them are semiconductors. Halogenation is comparable to alkalization in creating free carriers (Table 9.1). The metallic systems might be ferromagnetic or non-magnetic, depending on adatom concentration. The magnetic moment is in the range of $\sim 0.65 - 5.12 \; \mu_B$ for adatom concentra-

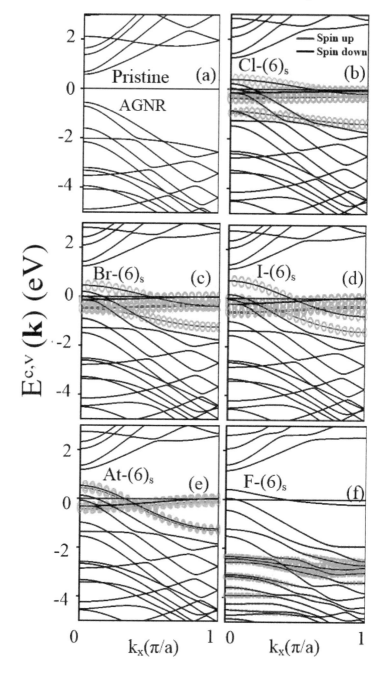

**FIGURE 10.7**
Band structures of $N_A = 10$ AGNRs for (a) H-terminated, (b) Cl-$(6)_s$, (c) Br-$(6)_s$, (d) I-$(6)_s$, (e) At-$(6)_s$, (f) F-$(6)_s$-adatom adsorptions. Green circles represent the contribution of halogen adatoms. The red and black curves denote the spin-split energy bands while the black and red numbers imply the top and bottom adsorption positions.

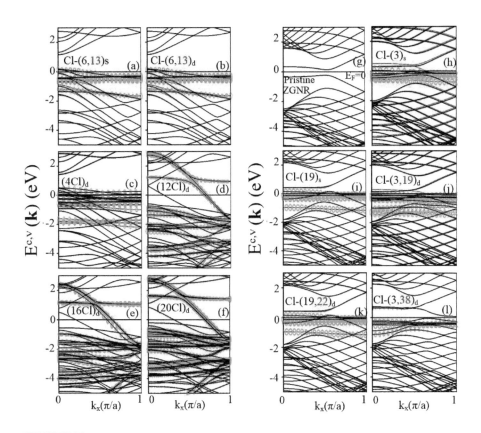

**FIGURE 10.8**

Band structures for: (a) Cl-(6,13)$_{s}$-, (b) Cl-(6,13)$_{d}$-, (c) (4Cl)$_{d}$-, (d) (12Cl)$_{d}$-, (e) (16Cl)$_{d}$-, and (f) (20Cl)$_{d}$-adsorbed $N_A = 10$ armchair systems; (g) H-terminated, (h) Cl-(3)$_{s}$, (i) Cl-(19)$_{s}$, (j) (3,19)$_{d}$, (k) (19,22)$_{d}$, and (l) (3,38)$_{d}$ $N_Z = 10$ zigzag systems. Green circles represent the contribution of Cl adatoms. The subscripts s and d, respectively, correspond to the single- and double-side adsorptions, while the black and red numbers imply the top and bottom adsorption positions.

tion lower than 60%; otherwise, it is vanishing. Non-magnetism indicates that the spin-dependent on-site Coulomb interactions [9] are fully suppressed by the strong X-X bondings at high concentration.

The Cl-induced spin configurations can create the diverse magnetic properties, namely FM and NM in armchair systems, and FM, NM, and AFM in zigzag systems. Armchair edges do not have the C-dependent spin configurations, so that only halogen adatoms dominate the spatial spin distributions. At concentrations < 60%, the spin-up configurations are induced by Cl adatoms, as shown in Figures 10.9(a)–10.9(e). Their densities grow with the increasing concentration, corresponding to the enhanced magnetic moment (Table 10.2). However, the strong Cl-Cl bondings under high concentrations are responsible for the vanishing magnetic configurations. On the other side, the C-dependent spin distributions at zigzag edges present the cooperative/competitive relation with Cl-induced spin configurations. When a single Cl exists at the top site of edge carbon (e.g., $(3)_s$ in Figure 10.9(g)), it generates the spin-up configuration and largely suppresses the spin-up and spin-down distributions due to the neighboring carbons (the H-terminated system in Figure 10.9(f)). Furthermore, this adatom also leads to the spin-up-dominated configuration at the lower edge. This ferromagnetic configuration is recovered to the anti-ferromagnetic one for two adatoms at the upper and lower edges (the $(3,38)_d$ adsorption in Figure 10.9(h)), since Cl- and C-induced spin configurations are identical. Apparently, the net magnetic moment is vanishing for two-edge absorption (Table 10.2). Such carbon-related edge spin configurations are hardly affected by the central adatom (the $(19)_s$ case in Figure 10.9(i)). However, an extra spin-up configuration appears there, leading to a new spin arrangement. Another spin distribution comes to exist, when two adatoms are, respectively, situated at edge and center (the $(3,19)_d$ adsorption in Figure 10.9(j)). The Cl-induced spin-up states dominate spin distribution, while the C-related edge spin distributions are thorough destroyed/largely suppresses.

In general, the fluorination and chlorination-related effects sharply contrast with each other in the essential properties, being attributed to the strength of orbital hybridizations in the X-C bonds and the halogen-induced spin configurations. All the halogen adatoms are situated at the top-site positions. However, the F-adatom adsorptions create the buckled structures, the shortest F-C bond length with the $(2p_x,2p_y,2p_z)$ multi-orbital hybridizations ($\sim$ 1.5 Å), the larger binding energy ($\sim$ −3.54 eV for F-$(6)_s$ adsorption), the (F,C)-co-dominated deeper valence bands clearly shown at various adsorptions, the p-type metals only under certain low-concentration distributions, the high charge difference between F and C, and five kinds of electronic and magnetic properties. The very strong F-C bondings destroy/seriously distort the π bondings, so that there is no simple relation between free hole densities and F concentrations. Furthermore, they greatly modify the $(2p_x,2p_y)$-dependent C-C and F-F bonds, leading to the merged van Hove singularities in DOSs associated with three kinds of chemical bonds. The F-adsorbed

**TABLE 10.2**

Binding energy, magnetic moment/magnetism, metal/energy gap, free hole density, X-C bond length, deviation height of C, and the first C-C bond length for halogenated $N_A = 10$ armchair and $N_Z = 10$ zigzag GNRs under various distributions and concentrations.

| GNRs | Systems | $E_b$ (eV) | $M_{tot}$ ($\mu$B) | $E_g$ (eV) | # of holes | X-C (Å) | C dev. (Å) | Nearest C-C (Å) |
|---|---|---|---|---|---|---|---|---|
| AGNR | H-terminated | - | 0/NM | 1.112 | 0 | - | 0 | 1.428 |
| | Cl-$(1)_s$ | -0.8232 | 0.74/FM | M | 1 | 3.345 | 0 | 1.430 |
| | Cl-$(6)_s$ | -0.8265 | 0.65/FM | M | 1 | 3.345 | 0 | 1.430 |
| | Cl-$(6,13)_s$ | -0.7932 | 1.28/FM | M | 1 | 3.357 | 0 | 1.433 |
| | Cl-$(6,13)_d$ | -0.8166 | 1.28/FM | M | 1 | 3.357 | 0 | 1.433 |
| | Cl-$(4F)_d$ | -0.9140 | 2.46/FM | M | 2 | 3.362 | 0 | 1.450 |
| | Cl-$(6F)_d$ | -0.9521 | 5.12/FM | M | 3 | 3.368 | 0 | 1.458 |
| | Cl-$(8F)_d$ | -1.0289 | 5.01/FM | M | 4 | 3.372 | 0 | 1.475 |
| | Cl-$(10F)_d$ | -0.9812 | 2.39/FM | M | 5 | 3.378 | 0 | 1.483 |
| | Cl-$(12F)_d$ | -0.9932 | 0/NM | M | 6 | 3.383 | 0 | 1.512 |
| | Cl-$(14F)_d$ | -0.9528 | 0/NM | M | 7 | 3.388 | 0 | 1.532 |
| | Cl-$(16F)_d$ | -0.9627 | 0/NM | M | 8 | 3.392 | 0 | 1.557 |
| | Cl-$(20F)_d$ | -0.9476 | 0/NM | M | 10 | 3.457 | 0 | 1.573 |
| | Cl-$(20F)_d$ | -0.6530 | 0/NM | 2.244 | 0 | 1.784 | 0.327 | 1.732 |
| | Br-(6) | -0.5496 | 0.69/FM | M | 1 | 3.615 | 0 | 1.429 |
| | Br-$(6,13)_d$ | -0.5021 | 1.41/FM | M | 1 | 3.684 | 0 | 1.430 |
| | I-(6) | -0.4585 | 0.52/FM | M | 1 | 3.750 | 0 | 1.430 |
| | At-(6) | -0.3398 | 0.19/FM | M | 1 | 4.107 | 0 | 1.430 |
| ZGNR | H-terminated | - | 0/AFM | 0.374 | 0 | - | 0 | 1.428 |
| | Cl-$(3)_s$ | -1.2795 | 0.29/FM | M | 1 | 3.009 | 0 | 1.432 |
| | Cl-$(19)_s$ | -1.3196 | 0.51/FM | M | 1 | 3.009 | 0 | 1.430 |
| | Cl-$(3,19)_d$ | -0.8631 | 0.67/FM | M | 1 | 3.002 | 0 | 1.430 |
| | Cl-$(19,22)_d$ | -0.8138 | 0.14/FM | M | 1 | 3.002 | 0 | 1.430 |
| | Cl-$(3,38)_d$ | -1.0535 | 0/AFM | M | 1 | 3.002 | 0 | 1.430 |
| | Br-$(19)_s$ | -0.8219 | 0.92/FM | M | 1 | 3.600 | 0 | 1.430 |
| | I-$(19)_s$ | -0.5981 | 1.04/FM | M | 1 | 3.823 | 0 | 1.430 |
| | At-$(19)_s$ | -0.3822 | 1.12/FM | M | 1 | 3.900 | 0 | 1.430 |

Note: NM, FM, and AFM correspond to non-magnetism, ferro-magnetism, and anti-ferro-magnetism, respectively.

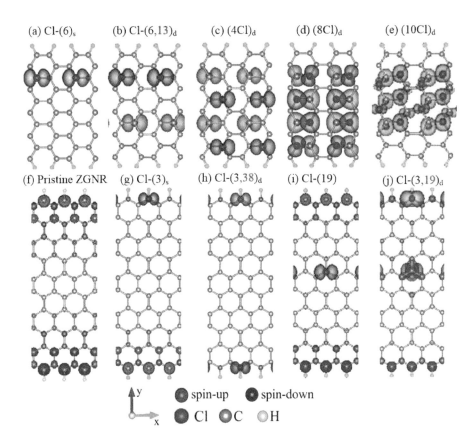

**FIGURE 10.9**
The spatial spin densities of Cl-adsorbed $N_A = 10$ armchair nanoribbons under various configurations: (a) Cl-$(6)_s$, (b) Cl-$(6,13)_d$, (c) $(4Cl)_d$, (d) $(8Cl)_d$, and (e) $(10Cl)_d$. Similar plots for $N_z = 10$ zigzag systems: (f) H-terminated, (g) Cl-$(3)_s$, (h) $(3,38)_d$, (i) Cl-$(19)_s$, and (j) $(3,19)_d$.

GNRs belong to FM and NM metals, NM semiconductors, and AFM semiconductors with/without spin splitting. Whether the F adatoms could generate the spin states of the neighboring carbons is very sensitive to distributions and concentrations. It is relatively difficult to observe the adatom-induced spin configurations, compared with other halogenated systems. On the other hand, the Cl-related systems present a planar honeycomb structure, the long X-C bond due to the weak single-orbital hybridization by their $p_z$ orbitals ($\sim 3.0 - 3.8$ Å), the smaller binding energy ($\sim -0.83$ eV for Cl-$(6)_s$ adsorption), the adatom-dominated energy bands with observable spin-split dispersions near the Fermi level at concentrations lower than 60%, the metallic behavior under any concentrations and distributions, three kinds of electronic and magnetic properties, and the concentration- and edge-dependent magnetic configurations. Two/one adatoms could provide one hole in a unit cell for concentrations above/below 10%. Such systems are FM, NM and AFM metals, in which the third kind only exhibits the spin-split behavior. The NM configuration concentrations come to exist at high concentrations, mainly owing to the full suppression of the rather strong X-X bond on the magnetic configuration. Moreover, the X-induced spin states and their serious competitions with the edge-C ones lead to the complicated spin distributions and thus diversify the spin-dependent energy bands and DOSs. In addition, there are also important differences between halogen- and alkali-adsorbed GNRs, covering the top-site/hollow-site adatom position, the p-type/n-type doping, and the existence/absence of magnetic configurations in armchair systems. That the alkali adatoms present the simple s-$2p_z$ hybridizations in X-C bonds and have no spin configurations is responsible for such differences.

The significant orbital hybridizations in chemical bonds could be fully comprehended by analyzing the spatial charge density. After the halogen adsorptions in armchair GNRs (Figures 10.10(b), 10.10(d), 10.10(f); 10.10(h)), the $\sigma$ bonding due to the planar $(2s,2p_x,2p_y)$ orbitals between two neighboring C atoms belong to very strong covalent bonds with high charge density, so they remain almost the same as the H-terminated GNR (Figure 10.10(a)). The $\pi$ bondings arising from the parallel $2p_z$ orbitals of C atoms could survive in a planar GNR for distinct concentrations, distributions, and edge structures, being the main mechanism of the metallic behavior. The changes on the $\pi$ bondings are identified through the charge variations between C and Cl atoms (the dashed red rectangles), and they are relatively easily observed at high concentrations (Figures 10.10(e) and 10.10(g)). These illustrate the single-orbital $3p_z$-$2p_z$ hybridization in the Cl-C bond. Specially, there exist the significant Cl-Cl bonds at the sufficiently high adatom concentration. The bonding evidences, which are associated with the $3p_z$-$3p_z$ and $(3p_x,3p_y)$-$(3p_x,3p_y)$ orbital hybridizations, are clearly presented in the charge density difference between two neighboring chlorines (the dashed pink and blue rectangles in Figures 10.10(e) and Figure 10.10(g)). The similar orbital hybridizations are also revealed in ZGNRs, as obviously indicated in Figures 10.10(h) and 10.10(k).

As mentioned above, chlorinated graphenes quite differ from fluorinated

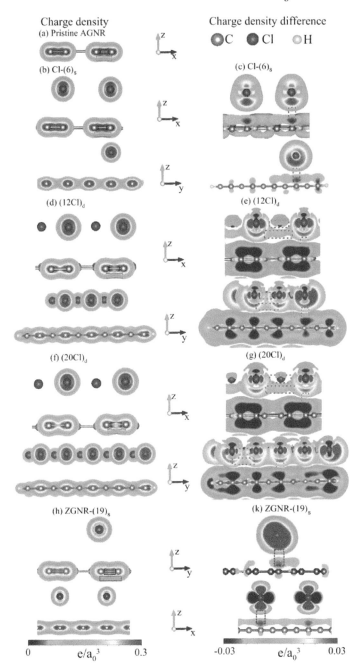

**FIGURE 10.10**
Spatial charge density situate at the left-hand side for (a) H-terminated, (b) Cl-(6)$_s$-, (d) (12Cl)$_d$-, (f) (20Cl)$_d$, and (h) ZGNR-Cl-(19)$_s$; charge density difference situate at the right-hand side for (c) Cl-(6)$_s$-, (e) (12Cl)$_d$-, (g) (20Cl)$_d$-, and (k) ZGNR-Cl-(19)$_s$.

systems in electronic properties, and so do the special structures in the orbital- and spin-projected DOSs (Figure 10.11). This further illustrates the orbital hybridizations and magnetic configurations. At low concentrations, energy spacing between the highest valence peak and the lowest conduction one present a blue shift, indicating the semiconductor-metal transition (Figures 10.11(b)–10.11(d)). The spin-up and spin-down components are different in the low-energy range of $(-1.5 \text{ eV} < E < 0.5 \text{ eV})$, being dominated by the $3p_z$ and $(3p_x, 3p_y)$ orbitals of Cl adatoms (the cyan and pink curves). Such result clearly demonstrates the Cl-created FM configuration, as revealed in the spin densities in Figures 10.9(a)–10.9(c). At high concentrations, e.g., $(12Cl)_d$ and $(20Cl)_d$ (Figures 10.11(e) and 10.11(f)), the spin degeneracy is recovered. Moreover, the Cl-related DOS covers a very wide range of $(-5 \text{ eV} < E < 3 \text{ eV})$ under the strong Cl-Cl bondings. Specifically, many peak structures of the Cl-$3p_z$ and C-$2p_z$ orbitals might appear at the same energies, mainly owing to the adsorption effects. On the other side, the low-lying DOS peak structures of ZGNRs are very sensitive to the adatom positions even at low concentrations. The H-terminated system has a pair of edge-C-dominated symmetric peaks across $E_F$ in the absence of spin splitting (purple circles in Figure 10.11(g)). These peaks become the spin-split structures after chlorination, as shown in Figures 10.11(h)–10.11(l). The Cl-dominated energy bands exhibit the similar peak structures. The total occupied states associated with such two kinds of peak structures could determine a specific magnetic configuration. The different numbers (same number) of spin-up and spin-down peaks at $E < 0$ corresponds to the FM (AFM) configuration, e.g., Cl-(3), Cl-(19), Cl-$(3,19)_d$ and Cl-$(19,22)_d$ cases in Figures 10.11(h)–10.11(k) (Cl-$(3,38)_d$ case in Figure 10.11(l)).

Up to now, the experiment verifications on the essential properties of halogenated GNRs are absent. The ARPES measurements, as done for the H-terminated systems [321, 331], could be utilized to examine the metallic/semiconducting band structures, the carbon- and adatom-dominated energy bands near the Fermi level, and the halogenation-dependent energy gaps. Furthermore, the spin-resolved ARPES [344] is useful to explore the spin-split electronic states. Both STS and spin-resolved STS are available in identifying the finite value at the Fermi level, energy gap, and spin-polarized peak structures [395, 331]. The transport experiments [397] could test the relation between hole density and halogen concentration, and the spin-polarized currents. Moreover, the magneto-optical experiments [112] are suitable for verifying the main features of the spin-split energy bands. Based on these experimental measurements, the critical orbital hybridizations in the chemical bondings and the specific spin configurations will be examined in detail, providing a full understanding on the halogenated GNR systems. It is worthy of mentioning that the high stability and remarkable properties of fluorinated GNRs are expected to make them become potential candidates in various application fields. For instance, edge-fluorinated graphene nanoplatelets or fluorinated graphene could be used as electrodes for lithium-based ion batter-

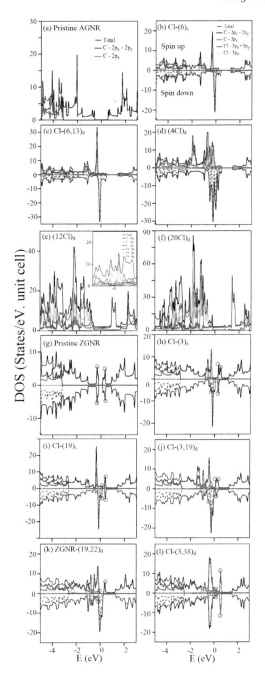

**FIGURE 10.11**
Orbital- and spin-projected DOSs for (a) H-terminated, (b) Cl-$(6)_s$, (c) Cl-$(6,13)_d$-, (d) $(4Cl)_d$-, (e) $(12Cl)_d$-, and (f)$(20Cl)_d$-adsorbed $N_A = 10$ armchair systems; (g) H-terminated, (h) Cl-$(3)_s$, (i) Cl-$(19)_s$, (j) $(3,19)_d$, (k) $(19,22)_d$, and (l) $(3,38)_d$ $N_Z = 10$ zigzag systems.

ies [143, 328]. Fluorine not only helps to activate sites for lithium storage but also facilitates the diffusion of Li ions during charging and discharging, leading to high-performance lithium batteries [328]. Specifically, the Cl-related GNRs might create the highest-density free holes and the spin-split energy bands, so that they are very helpful in developing the spintronic nanodevices [152, 156, 366].

# 11

## Metal-adsorbed graphene nanoribbons

Up to now, have been a lot of studies associated with aluminium-based bat-teries. On the graphite cathode [206], the predominant $AlCl_4$ molecules are, respectively, intercalated and de-intercalated between graphitic layers during charge and discharge processes. On the aluminium anode, the metallic Al and $AlCl_4$ are transformed into $Al_2Cl_7^-$ during discharging, and the reverse chemical reactions happen under charging. Al atoms are deduced to play a critical role in the giant enhancement of current densities. The substitution of lithium with aluminum anode, which is characterized by the lower cost and higher abundance, is one of the most widely studied methods to reduce the cost of electrochemical storage systems and enable the long-term sustainabil-ity [92]. On the theoretical side, the first-principles calculations on Al-doped graphenes predict that such systems are suitable in applying to environment and energy engineerings, such as toxic gases senors/detectors [13, 73], and a potential hydrogen storage [12]. Moreover, the Al-absorbed 2D monolayer graphene shows the maximum adatom concentration corresponding to a $2 \times 2$ enlarged unit cell [216]. The optimal position is the hollow site with the height of 2.00 $\overset{\circ}{A}$ and the Al-C bond length of 2.54 $\overset{\circ}{A}$. This system exhibits the red shift of the Fermi level, in which the well-known Dirac cone, with the linear energy dispersions, is preserved or only slightly distorted. That is to say, the low-lying energy bands are dominated by the original $\pi$ bonding of carbons. The free electron densities arising from the $n$-type dopings are very high and quite close to those observed in alkali-adsorbed graphenes. There no mag-netic properties after aluminum adsorption. However, there exist the signif-icant multi-orbital hybridizations in Al-C bonds, namely, the $(3s, 3p_x, 3p_y)$ and $2p_z$ orbitals. The main reason might be that each Al adatom has three outmost valence electrons. They are revealed as the (Al,C)-co-induced valence and conduction bands out of the Dirac-cone structure and the merged peaks in the orbital-decomposed DOSs. The finite-size effects and the edge structures are expected to greatly modify the electronic and magnetic properties.

In recent years, various doping methods for epitaxial few-layer graphenes are realized by adatom adsorption and substitution. The previous experimen-tal and theoretical study on the titanium-absorbed graphene has confirmed the greatly modified band structures by using the high-resolution ARPES mea-surements [66]. By depositing a low concentration of Ti atoms on epitaxial graphene supported by SiC, the electronic band structure of graphene could be significantly engineered. The titanium low coverage is explored in experimental measurements, since that Ti adatoms are easy to from nanoclusters rather than isolated adatoms [54, 236]. Such adatoms on graphene surface are predicted to have large cohesive energies, indicating there exists the critical orbital hybridization in the Ti-C bonds. The first-principles calculations reveal that a very strong hybridization occurs primarily between the vertical Ti-$3d$ and C-$2p_z$ orbitals as a result of the spatial overlap and symmetry matching. This is responsible for the high doping efficiency and the readily tunable carrier density. This Ti-induced hybridization is adsorbate-specific and creates major consequences for the efficient doping as well as the potential applications towards adsorbate-induced modification of carrier transport in graphene. For example, the titanium-covered graphene is proved to be very efficient in hydrogen adsorption and desorption [236], but not the formation of $H_2$ molecules.

Nowadays, the bismuth-related systems are one of the most widely studied materials, because the dimensionality could enrich the fundamental properties. The bulk bismuth, which possesses a rhombohedral symmetry, is a semimetal with a long Fermi wavelength and small effective electron mass [127]. Its surface states belong to a Dirac fermion gas [190]. The three bismuth surfaces: Bi(111), Bi(100), and Bi(110) have a higher free carrier density at the Fermi level compared to those of the bulk system [127]. The Bi thin film, without strong chemical reactions by $O_2$ on surfaces, appears to be stable up to about 600 K [36]. Recently, the 2D few-layer bismuthenes are epitaxially grown the 3D $Bi_2Te_3$(111)/$Bi_2Se_3$(111)/Si(111) substrates [123, 122, 375, 358, 124]. Also they have been successfully obtained from the mechanical exfoliation [297]. Monolayer bismuthene is predicted to exhibit the rich and unique magnet-electronic properties, mainly owing to the cooperation among the specific geometric symmetry, the multi-orbital hybridizations in Bi-Bi bonds, the significant spin-orbital couplings, the magnetic field and the electric field [68]. Moreover, the 1D bismuth nanowires exhibit narrow band gaps due to significant quantum confinement effects [32]. The above-mentioned 3D−1D bismuth systems have the significant $sp^3$ bondings and the strong spin-orbital couplings, in which the former mainly come from the $(6s, 6p_x, 6p_y, 6p_z)$ orbitals of Bi atoms. The $sp^3$ tight-binding models, with the spin-orbital interactions, are successful in understanding the magneto-electronic properties of the semi-metallic 3D rhombohedral bismuth [222], and the semi-conducting 2D bismuthene [142]. In addition, bismuth systems are explored in detail for the fields of environmental engineering, biochemistry, and energy engineering, such as Bi-based nanoelectrode arrays in detecting heavy metals [353], a polycrystalline bismuth oxide film as a biosensor [309], and bismuth oxide on nickel foam available for the anode of lithium battery [195].

Recently, bismuth adatoms on monolayer graphene supported by a 4H-SiC(0001) substrate have been obviously observed at room temperature [64, 63]. The corrugated substrate and buffer graphene layer are clearly identified from the STM experiments, in which a large-scale hexagonal array of Bi adatoms appears at room temperature. Such adatoms could form the triangular and rectangular nanoclusters of a uniform size by the further annealing process. Moreover, the STS measurements of the dI/dV spectra confirm the existence of the Dirac-cone structure by the $V$-shape differential conductance, indicating the blue shift of the Fermi level (the $p$-type doping) and the creation of free conduction electrons. In addition, the Bi-related structures in DOS are revealed at deeper energy. These important results shed light on controlling the nucleations of the isolated adatoms and nanostructures on graphene surfaces. They are successfully simulated by the six-layered substrate, the corrugated buffer layer, and the slightly deformed monolayer graphene using the first-rinciples calculations [213]. The Bi adatom arrangements are thoroughly in-vetigated by analyzing the ground state energies, bismuth adsorption energies, and BiVBi interactions of energies under the different heights, inter-adatom distances, adsorption sites, and hexagonal positions. A hexagonal array of Bi adatoms is dominated by the van der Waals interactions between the buffer layer and monolayer graphene. An increase in temperature could overcome a $\sim 50$ meV energy barrier and generate the unusual triangular and rectangular nanoclusters. The most stable and metastable structures are consistent with the experimental observations. There also exist some theoretical studies on the geometric structures and energy bands of Bi-adsorbed and Bi-intercalated graphenes [8, 129]. The former is conducted on monolayer graphene with-out simulation of the substrate and the buffer graphene layer, and thus the deformed graphene surface structure might be unreliable [8]. The latter is eval-uated for Bi and/or Sb as a buffer layer above the four-layer SiC substrate, which lead to an energetically unfavorable environment for the metal adatoms to be adsorbed on the graphene sheet [129].

The metal atoms, the alkali ones excepted, might provide the multiple outermost orbitals for the multi-orbital hybridizations with the out-of-plane $\pi$ bondings on the honeycomb lattice. This will dominate the fundamental properties of Al-, Ti- and Bi-adsorbed graphene nanoribbons, in which they are explored thoroughly by using the first-principles calculations. The prin-ciple focuses are the adatom-dependent binding energies, the adatom-carbon lengths, the optimal position, the maximum adatom concentrations, the free electron density transferred per adatom, the adatom-related valence and con-duction bands, the various van Hove singularities in DOSs, the transition-metal-induced magnetic properties, and the significant competitions of the zigzag edge carbons and the metal/transition metal adatoms in spin configu-rations. The distinct chemical bondings are clearly identified from three kinds of metal adatoms under the delicate physical quantities. The important dif-ferences between Al-/Ti-/Bi- and alkali-adsorbed graphene nanoribbons will be discussed in detail, covering band structures, relation of conduction elec-tron density and adatom concentration, spatial charge distributions, orbital-decomposed DOSs, and magnetic configurations and moments.

## 11.1  Al

The Al-adsorbed graphene nanoribbon remains the planar structure with a non-uniform honeycomb lattice, as revealed in Figures 11.1(a)–11.1(f). This clearly indicates the well-behaved $\sigma$ bondings of carbon atoms, being hardly affected by the Al adsorptions. The optimal positions might correspond to the hollow sites if the adatoms are far away from the boundaries, i.e., the $(x, y)$-plane projections are the centers of hexagon lattice in the absence of $y$-shift (Table 11.1). But when the Al adatoms are located near the armchair and zigzag boundaries (Figures 11.1(a)–11.1(f)), the obvious shifts are revealed along the transverse direction toward the edges, e.g., $\sim 0.25 - 0.32$ Å for the 1, 2, 7, and 8 positions in $N_A = 10$ armchair system and $\sim 0.06 - 0.17$ Å for the 1, 3, and 5 positions in $N_z = 10$ zigzag one. The height of Al adatom is $2.07 - 2.21/2.00 - 2.14$ Å for armchair/zigzag systems, in which the maximum value is associated with the nanoribbon center. That is, the Al adatoms are relatively low under the effect of the edge C-H bonds. As to the highest concentration, the stable structure is associated with the double-side adsorption (not the single-side adsorption) of four Al adatoms in $N_A = 10$ armchair system, and it is related to the similar adsorption of seven Al adatoms in $N_z = 10$ zigzag one. This result does not exceed the upper limit of 25% in Al-adsorbed monolayer graphene [216]. There exist the unusual geometric structures under the maximum concentration, as obviously indicated in Figures 11.1(d) and 11.1(f). The optimal position is dramatically transferred to the bridge side, clearly illustrating the complex competitions/cooperations among the C-C, Al-C, Al-Al and H-C chemical bondings. This means that the two Al adatoms need to have the sufficiently long distance to achieve the stable structure. The high-resolution STM and TEM could be utilized to verify the first-principles predictions. After Al adsorption, the similar 1D graphene plane suggests that only the $2p_z$ orbitals of carbons have significant chemical bondings with the half occupied three kinds of orbitals $(3s, 3p_x, 3p_y)$, and the $\sigma$ bondings of $(2s, 2p_x, 2p_y)$ almost keep the same. This will be explored thoroughly.

The electronic structures exhibit the drastic changes in all graphene nanoribbons during the aluminization, as clearly shown in Figures 11.2 and 11.3. In general, the Fermi level is shifted from the center of energy gap to the conduction bands. Furthermore, the energy bands very close to $E_F$ are mainly determined by carbon atoms, indicating the free electrons due to the distorted $\pi$ bondings and the charge transfer from carbon atoms to Al adatoms. For armchair systems, the energy spacing between the first pair of valence and conduction bands is related to $k_x \neq 0$ or $k_x = 0$, in which the former might depend on whether the Al adatoms are situated at armchair edges (the $(1)_s$ case in Figure 11.2(a)). Specifically, the Al adatoms make important contributions to certain conduction and valence bands in the range of $0 \leq E^c \leq 2.0$ eV and $-2.5$ eV$\leq E^v \leq -4.3$ eV, in which the $-2.5$ eV$\leq E^v \leq -3.5$ eV

**TABLE 11.1**
Binding energy, adatom height, adatom y-shift, magnetic moment/magnetism
for Al-adsorbed $N_A = 10$ armchair and $N_Z = 10$ zigzag GNRs under various
distributions and concentrations.

| | Eb (eV) | Height (Å) | Adatom y-shift (Å) | M ($\mu$B) |
|---|---|---|---|---|
| AGNR;Al$(1)_s$ | -0.641 | 2.071 | 0.319 | 0 |
| $(5)_s$ | -0.639 | 2.207 | x | 0 |
| $(1,8)_s$ | -0.636 | 2.063 | 0.252 | 0 |
| $(1,8)_d$ | -0.636 | 2.067 | 0.289 | 0 |
| $(2,7)_s$ | -0.453 | 2.077 | x | 0 |
| $(2,7)_d$ | -1.969 | 2.065 | x | 0 |
| $(1,2,7,8)_d$ | -0.714 | 2.153 | 0.144; 1.231 | 0 |
| ZGNR;Al$(1)_s$ | -1.606 | 2.049 | 0.172 | 0.56/FM |
| $(3)_s$ | -1.229 | 1.996 | 0.072 | 0.52/FM |
| $(5)_s$ | -1.074 | 2.076 | 0.064 | 0.18/FM |
| $(9)_s$ | -1.142 | 2.135 | x | 0/AFM |
| $(1,17)_s$ | -1.789 | 1.999 | 0.317 | 0 |

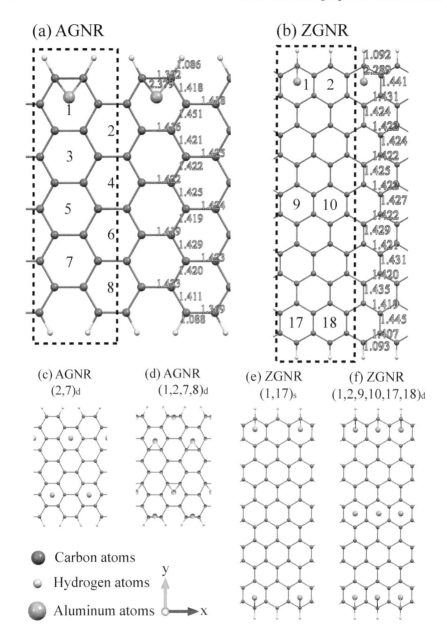

**FIGURE 11.1**

The non-uniform optimal geometric structures for the Al-adsorbed $N_A = 10$ armchair graphene nanoribbons under the initial adatom positions: (a) $(1)_s$ (c) $(2,7)_s$ and (d) $(1,2,7,8)_d$; those of the $N_Z = 10$ systems with the similar positions: (b) $(1)_s$, (e) $(1,17)_d$ and (f) $(1,2,9,10,17,18)_d$.

valence states are weakly dispersive and almost doubly degenerate. The Al adsorption could create the extra band-edge states arising from the subband hybridizations. Energy bands are very sensitive to the changes in the position and concentration of adatoms (Figures 11.2(a)–11.2(d)), but not the single- or double-side adsorption (Figures 11.2(c) and 11.2(d)).

The free conduction electrons provided by per Al adatom in a unit cell are worthy of closer examinations. With the increasing Al concentration, there are more carbon-dominated conduction bands intersecting with the Fermi level, as obviously revealed in Figures 11.2(c) and 11.2(d) for aluminization with two adatoms. By the addition of the Fermi momenta, the 1D linear carrier density becomes double, regardless of the single- or double-side adsorption, and the adatom positions. The linear relation between conduction electron density and adatom concentration is roughly achieved under the specific three Al adsorptions, e.g., the $(1, 5, 8)$ and $(1, 2, 8)$ cases. However, this simple relation is absent under the maximum adsorption concentration. The four-Al case might correspond to a indirect-gap semiconductor with a very small $E_g$, e.g, $E_g = 0.04$ eV under the $(1,2,7,8)_d$ adsorption.

Energy bands strongly depend on the edge structures; that is. there are certain important differences in zigzag and armchair graphene nanoribbons. For the single adatom adsorption on a zigzag system, whether the low-lying energy bands exhibit the spin spitting is very sensitive to the position of Al, as clearly indicated in Figures 11.3(a)–11.3(d). There exist the drastic changes in energy bands, especially for electronic states near $k_x = 0$ and $\pm 2/3$. A pair of partially flat and spin-degenerate valence and conduction bands, which, respectively, cross the Fermi level and are almost symmetric about $E_F = 0$, dramtically change the energy dispersions, possess the Fermi momenta in the conduction bands, and even present the spin spitting. When the Al adatom is not located at the nanoribbon center, the spin-spilt metallic behavior comes to exist, as shown in Figures 11.3(a)–11.3(c). Specially, the spin-split energies arising from the partially flat bands even reach $\sim 0.5 - 0.7$ eV close to the $k_x = 0$ state. The occupied state number is different in the spin-up- and spin-down-related energy bands; that is, such zigzag systems belong to the FM metal, in which the net magnetic moment is 0.56 $\mu_B$. per unit cell. On the other hand, the central single-Al adsorption cannot create the spin splitting and thus preserve the original AFM spin configuration ((the $(9)_s$ case in Figure 11.3(d)). Band structures become more complicated in the increase of adatom concentration. When there are two Al adatoms close to the distinct boundaries, the magnetic configuration are fully absent. For example, the $(1.17)$ adsorption in Figure 11.3(e) is a non-magnetic metal, in which the valence and conduction energy bands are doubly degenerate near the $k_x = 0$ state because of the merged neighboring bands.

The Al-adsorbed zigzag graphene nanoribbons exhibit the diverse magnetic configurations, being different from/similar to the spin arrangements of the alkali-adsorbed ones. A single Al close to the upper zigzag boundary will partially destroy the spin configutation due to the zigzag edge carbons,

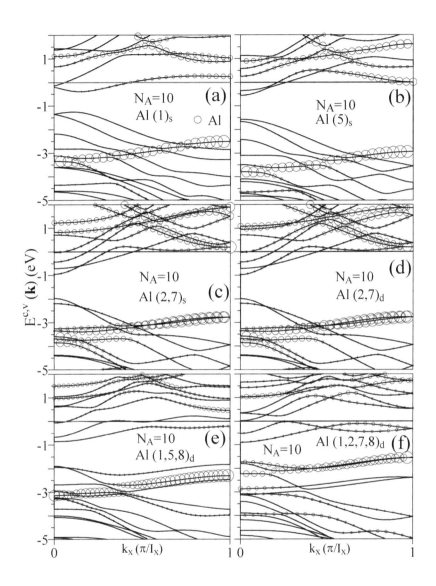

**FIGURE 11.2**
Band structures of the Al-adsorbed $N_A = 10$ armchair graphene nanoribbons
for the various adsorptions: (a) $(1)_s$, (b) $(5)_s$, (c) $(2,7)_s$, (d) $(2,7)_d$, (e) $(1,5,8)_d$,
and (f) $(1,2,7,8)_d$.

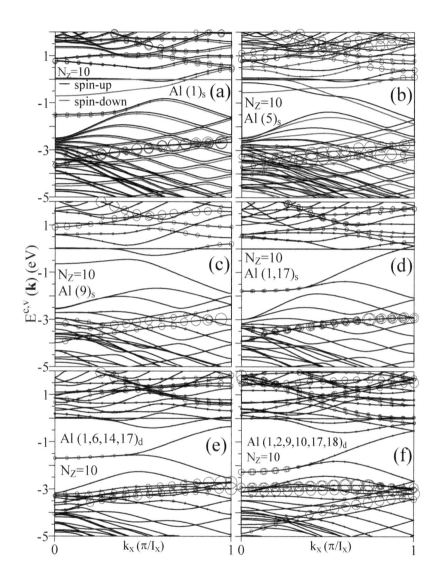

**FIGURE 11.3**
Similar plot as Figure 11.2, but displayed for the $N_Z = 10$ zigzag systems under the various cases: (a) $(1)_s$, (b) $(5)_s$, (c) $(9)_s$, (d) $(1,17)_s$, (e) $(1,6,14,17)_d$, and (f) $(1,2,9,10,17,18)_d$.

depending on the adatom positions, as clearly revealed in Figures 11.4(a)–11.4(c). The lower zigzag edge preserves the spin-up-dominated configuration (red balls) almost identical to the original case. However, the spin-down arrangement (blue balls) near the upper edge is fully/partially annihilated under the $(1)_s$ and $(3)_s/(5)_s$ cases, and the spin-up distribution might be drastically changed from edge to center. Specially, the spin-up configuration in the whole zigzag nanoribbon has been created by the $(1)_s$ and $(3)_s$ adsorptions (Figures 11.4(a) and 11.4(b)); that is, the edge and non-edge carbons possess the magnetic configurations simultaneously. This is absent in the alkali-adsorption systems (Figures 9.7(b) and 9.7(c)), It should have the maximum magnetic moment for the Al adatom near the zigzag edge. On the other hand, one central Al adatom hardly affects the original AFM spin distribution, being similar to the alkali case (Figure 9.7(e)). The magnetic properties thorough vanish under two Al adatoms near the separate zigzag edges, as observed in alkali-adsorbed systems (Fig 9.7(f)). The above-mentioned features of magnetic properties clearly show the very strong competition/cooperation between the multi-orbital hybridizations of Al-C bonds and the spin sates due to zigzag edge carbons in the total ground state energy.

The carrier density and the variation of carrier density can provide very useful informations about the orbital bondings, energy bands, and charge transfer. The former directly reveals the bonding strength of C-C, Al-C, and Al-Al bonds, as illustrated in Figures 11.6(a)–11.6(d). After the Al adatom adsorptions, all the C-C bonds possess the strong covalent $\sigma$ bonds (blue rectangle) and the somewhat weaken $\pi$ bonds simultaneously (red rectangle). The former almost keep the same; furthermore, the $\pi$ bonding also belongs to the extended state in a 1D system except that it is seriously distorted under the maximum-concentration case (Figure 11.2(f)). These are responsible for the carbon-dominated low-lying energy bands with the metallic or semiconducting behavior (Figure 11.5). In general, the $3s$-orbital electrons (black triangle) are redistributed between Al and the six nearest C atoms, revealing a significant hybridization with the $2p_z$ orbitals (yellow rectangle; green ring in the inset). The spatial charge distribution of $(3p_x, 3p_y)$ orbitals is extended from the light blue ring (inset) near Al to the red ring between C and Al atoms. It is noticed that these orbitals make less contribution to the chemical bonding. The multi-orbital hybridizations of $3s$ and $(3p_x, 3p-y)$ are, respectively, related to the Al-dominated valence and conduction bands (Figure 11.2). The electrons of Al adatom move to the top and bottom of the non-nearest C atoms (red region within a black rectangle), inducing free electrons in conduction bands. The higher the Al-concentration is, the more the electrons are transferred (Figure 11.4(f)).

The Al-adatom adsorptions can create the metallic DOSs in graphene nanoribbons except for the maximum concentration (Figures 11.6(a)–11.6(h)), while the main characteristics of van Hove singularities are in sharp contrast with those of H-terminated pristine systems (Figure 3.6 and 3.12) and alkali-adsorbed cases (Figures 9.5 and 9.8). The low-energy DOSs are dominated by

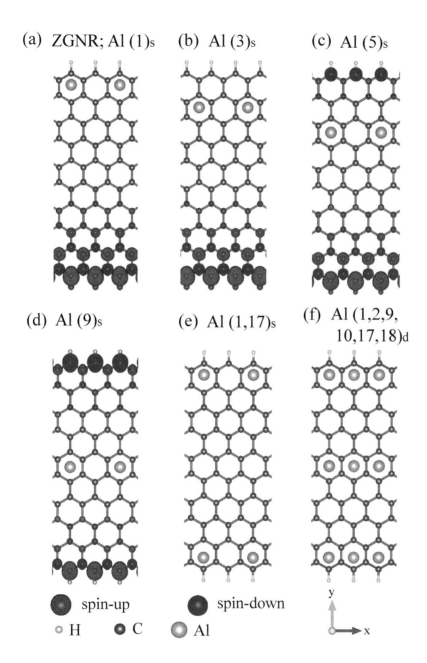

(a) ZGNR; Al $(1)_s$  (b) Al $(3)_s$  (c) Al $(5)_s$

(d) Al $(9)_s$  (e) Al $(1,17)_s$  (f) Al $(1,2,9, 10,17,18)_d$

spin-up  spin-down

○ H  ● C  ○ Al

**FIGURE 11.4**

The spatial spin distributions of the Bi-adsorbed $N_Z = 10$ zigzag systems due to the various adsorptions: (a) $(1)_s$, (b) $(3)_s$, (c) $(5)_s$, (d) $(9)_s$, (e) $(1,17)_d$, and (f) $(1, 2, 9, 10, 17, 18)_d$. They are shown on the $x - y$ plane.

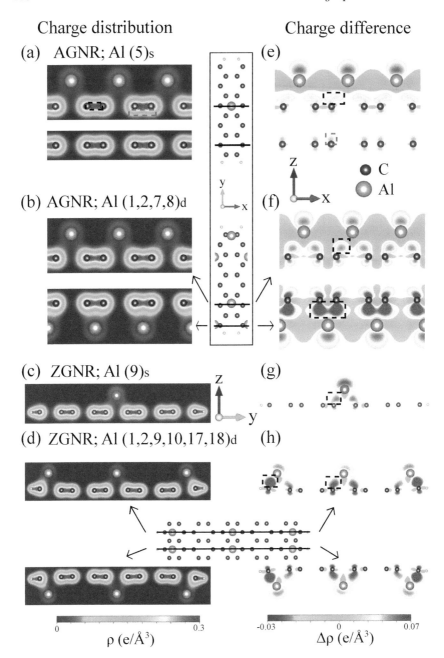

**FIGURE 11.5**
The spatial charge distributions and the differences after chemisorptions.
$\rho$'s/$\Delta\rho$ on the $x-z$ plane are shown for the $N_A = 10$ armchair systems with
Bi adatoms at (a)/(e) $(5)_s$ and (b)/(f) $(1,2,7,8)_d$; those on the $y-z$ plane
for the $N_Z = 10$ zigzag systems under the Bi-adsorptions: (c)/(g) $(9)_s$ and
(d)/(h) $(1,2,9,10,17,18)_d$.

the $2p_z$ orbitals of carbons, being consistent with the atom-dominated band structures (Figures 11.2 and 11.3). This feature clearly illustrates that the $\pi$ bondings are distorted only near the Al adatoms, and they behave as the normal extended states at others. For armchair systems (Figures 11.6(a)–11.6(d)), there exist the specific zero DOSs below the Fermi level in the range of $-2$ eV$\le E^v \le -1$ eV, belonging to the initial valence and conduction bands. However, the similar DOSs have a finite value in zigzag systems (Figures 11.6(e)–11.6(h)). Concerning the clear evidences of the multi-orbital hybridizations in Al-C bonds, they show as the merged peaks in the range of $E^v < -1.5$ from the Al-3$s$ orbital and the C-2$p_z$ orbital, and those for $E^c > -0.5$ eV due to the Al-($3p_x$,$3p_y$) orbitals and the C-2$p_z$ orbital. Energy bandwidths of the 3$s$ and ($3p_x + 3p_y$) orbitals grows gradually in the increment of Al concentrations, indicating two kinds of orbital hybridizations in Al-Al bonds. It should be noticed that the conduction states below the Fermi level about $\sim 1$ eV also make inportnt contributions to the essential properties. The $3p_z$ orbitals of Al adatoms hardly contribute to the significant chemical adsorption, since each adatom only has three occupied orbitals in the outermost ones, in which it might possess two states in the 3$s$ orbitals and one state in the ($3p_x$, $3p_y$) orbitals (from the covered areas by the 3$s$ and $3p_x + 3p_y$ orbitals, respectively). Specifically, it might be difficult to identify the special peaks coming from the low-lying partially flat bands, and the spin-up- and spin-down-split DOSs are obviously revealed in certain adsoption configurations of zigzag graphene nanoribbons (Figs, 11.5(e) and 11.5(h)). The FM metals are useful in exploring the spintronic transports and the spin-related applications.

## 11.2 Ti

The Ti-adatom chemisorptions on 1D graphene nanoribbons could create the optimal geometric structures similar to 2D graphene systems except for the non-uniform band lengths. Figures 11.8(a)-11.8(f) clearly show the optimal hollow sites without any shifts relative to the $(x, y)$-projection centers under all the adsorption configurations. This is in sharp contrast with the significant shifts of the Al and Bi metal adsorptions (Sections 11.1 and 11.3). The Ti heights keep in the short range of $\sim 1.50 - 2.10$, leading to the largest binding energies ($\sim -2.5 - 3.5$ eV) among the (Ti,Al,Bi) metal adatoms. It is relatively easy to induce the Ti chemisorptions, compared with the Al and Bi metal adatoms. The former could reach/present the higher adatom concentrations. Specifically, the buckling structures, might be revealed for the highest-concentration adsorption in armcahir/zigzag systems, e.g., $(1,2,3,4,5,6,7,8)_d/(1,2,5,6,9,10,13,14,17,18)_d$ in Figure 11.8(c)/Figure 11.8(f). The main features of optimal structures strongly suggest that the $2p_z$ and

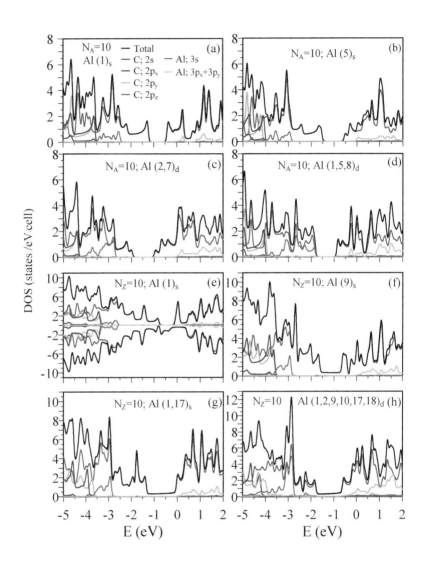

**FIGURE 11.6**
The orbital- and spin-projected DOSs for the $N_A = 10$ armchair systems under the various adsorptions: (a) $(1)_s$, (b) $(5)_s$, (c) $(2,7)_d$, (d) $(1,5,8)_d$, and those for the $N_Z = 10$ systems with the adatom positions: (e) $(1)_s$, (f) $(9)_s$, (g) $(1,17)_s$, and (h) $(1,2,9,10,17,18)_d$.

**TABLE 11.2**
Binding energy, adatom height, adatom y-shift, magnetic moment/magnetism
for Ti-adsorbed $N_A = 10$ armchair and $N_Z = 10$ zigzag GNRs under various
distributions and concentrations.

| | Eb (eV) | Height (Å) | Adatom y-shift (Å) | M ($\mu$B) |
|---|---|---|---|---|
| AGNR;Ti(1)$_s$ | -2.828 | 1.777 | x | 1.28/FM |
| (5)$_s$ | -2.591 | 1.784 | x | 1.52/FM |
| (1,8)$_s$ | -2.790 | 1.769 | x | 2.27/FM |
| (2,7)$_s$ | -2.542 | 1.838 | x | 2.91/FM |
| (2,7)$_d$ | -2.545 | 1.811 | x | 2.89/FM |
| (1,2,7,8)$_s$ | -3.783 | 1.729/1.859 | x | 1.46/FM |
| (1,2,7,8)$_d$ | -2.854 | 1.506/2.109 | x | 4.22/FM |
| (1,2,4,5,7,8)$_d$ | -3.376 | 1.779/1.919 | x | 4.16/FM |
| (1,2,3,4,5,6,7,8)$_d$ | -3.454 | 1.822/1.985 | x | 5.93/FM |
| ZGNR;Ti(1)$_s$ | -2.774 | 1.711 | x | 1.46/FM |
| (5)$_s$ | -2.273 | 1.728 | x | 1.52/FM |
| (9)$_s$ | -1.991 | 1.629 | x | 0/AFM |
| (1,17)$_s$ | -2.564 | 1.670 | x | 0/AFM |
| (1,9,17)$_s$ | -2.207 | 1.618 | x | 0/AFM |
| (1,6,14,18)$_d$ | -2.354 | 1.588/1.607 | x | 0/AFM |
| (1,2,5,6,9,10,13,14,17,18)$_d$ | -2.606 | 1.665/1.718 | x | 0/AFM |

$(2s, 2p_x, 2p_y)$ orbitals of carbon atoms, respectively, make important and minor contributions to the Ti-C bonds.

Any Ti-adsorbed graphene nanoribbons belong to the unusual metals, as
clearly indicated in Figures 11.8 and 11.9. There are a lot of energy bands
crossing the Fermi level, in which they mainly come from both Ti adatoms
and carbon atoms. Furthermore, the Ti-dominated energy bands might be
thoroughly occupied or unoccupied. For armchair systems, all of them exhibit
the spin-split energy bands, especially for those near $E_F$. Apparently, they
correspond to the FM spin configurations, and the net magnetic moments
due to the Ti adsorbates are sensitive to the adatom distribution and concentration (Table 11.2). Under the single- and two-adatom adsorptions (Figures
11.8(a)–11.8(d)), the Ti-dependent electronic structures lie in the range of $-1$
eV$\leq E^{c,v} \leq 3$ eV, and the host atoms fully determine the energy bands below
it. The increasing adatom concentration, respectively, makes major and minor
contributions to these two different ranges of energy bands. The Ti adatoms
hardly contribute to the deep valence states, indicating the weak chemical
hybridizations from the $(2s, 2p_x, 2p_y)$ orbitals in the Ti-C bonds.

The Ti chemisorptions could induce the diversified band structures be-

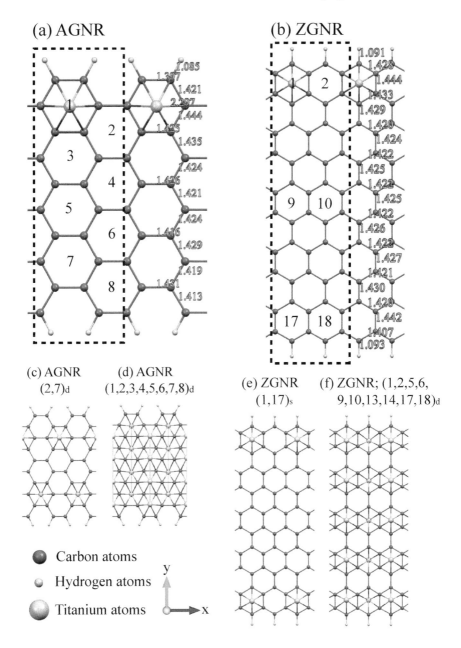

**FIGURE 11.7**

The optimal geometric structures for the Ti-adsorbed $N_A = 10$ armchair systems under the adatom positions: (a) $(1)_s$, (c) $(1,8)_s$, and (d) $(1,2,3,4,5,6,7,8)_d$; those of the $N_z = 10$ zigzag systems with the similar positions: (b) $(1)_s$, (e) $(1,17)_d$, and (f) $(1,2,5,6,9,10,13,14,17,18)_d$.

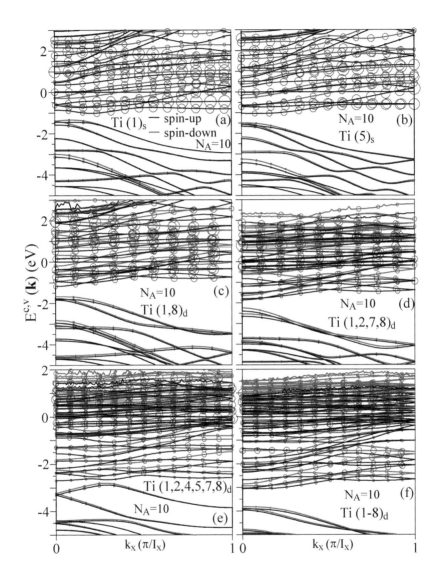

**FIGURE 11.8**
Band structures of Ti-adsorbed $N_A = 10$ armchair nanoribbons under the various adsorptions: (a) $(1)_s$, (b) $(5)_s$, (c) $(1,8)_d$, (d) $(1,2,7,8)_d$, (e) $(1,2,4,5,7,8)_d$, and (f) $(1\text{-}8)_d$.

cause of the edge structures. There are certain important differences between the zigzag and armchair systems. As for the former, the partially flat bands across the Fermi level only survive under the single-adatom central adsorption (Figure 11.9(b)). However, most of chemisorption cases, being shown in Figures 11.9(a) and 11.9(c)–(f), lead to the dramatic changes in the low-lying energy bands because of the significant Ti-edge-C bondings; that is, the edge-C-dominated partially flat energy dispersions are thoroughly absent. The zigzag systems might be FM, AFM and NM metals, as clearly revealed in Figures 11.9(a)–11.9(f). in which the latter two need to be further examined from the spatial spin distributions. Apparently, this is purely due to the strong competition of edge carbon atoms and Ti adatoms. The first kind of spin configuration, which is characterized by the distinct occupied states in the spin-up and spin-down energy bands, corresponds to the non-symmetric adatom distributions (Figure 11.9(a)). Furthermore, the second and third kinds are closely associated with the symmetric ones (Figures 11.9(b)–11.9(f)). They exhibit the spin-degenerate band structures.

There exist the diverse various FM and AFM spin distributions. Any armchair systems have the distinct FM configuratons, being  sensitive to the single- and double-side absorptions and concentrations. The single-adatom chemisorptions, e.g., $(1)_s$ and $(5)_s$ (Figures 11.10(a) and Figure 11.10(b)), could create the comparable magnetic moments ($\sim 1.1 - 1.2\mu B$ in Table 11.2) and slightly induce the identical spin arrangement of the neighboring carbon atoms. In general, the strength of magnetic response is proportional to the Ti concentration under the specific double-side adsorptions. For example, the red spin-related volumes grow with the inceeasing concentrations, $(1)_s$, $(1,8)_d$, $(1,2,7,8)_d$; $(1-8)_d$ respectively, in Figures 11.10(a), 11.10(d), 11.10(f); 11.10(g), and so do the net magnetic moments. However, for the single-side chemisorptions, the neighboring Ti adatoms might compete with each other in spin states and thus create the abnormal spin arrangements, e.g., the $(1,2,7,8)_s$ case with the smaller magnetic moment in Figure 11.10(e).

The unusual competitions between zigzag carbons and Ti adatoms could lead to the diverse spin distributions, being strongly associated with the partial or full spin suppressions of the former and the latter. The FM and AFM configurations, respectively,  correspond to the asymmetric and symmetric adatom distributions in zigzag systems. The single-adatom chemisorptions create the FM configuration ($(1)_s$ and $(5)_s$ in Figures 11.11(a) and 11.11(b)) except for the symmetric case ($(9)_s$ in Figure 11.11(c)). The guest adatom, which is located at (near) the zigzag boundary, has a wide spin-up distribution (red balls) and full destroys the spin-up-dominated arrangement from the neighboring carbons (even changes into the spin-down-determined configuration). Furthermore, there exists the spin-down dominant distribution on the other boundary (blue balls). However, the spin state of the Ti adsorbate is absent under the symmetric adatom distribution, in which the pristine AFM configuration across the ribbon center is reduced, but remains similar. The absence of magnetism of Ti adatom might arise from the symmetric chemical

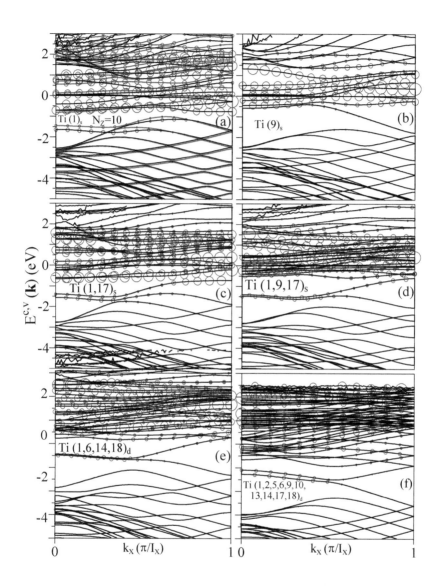

**FIGURE 11.9**
Similar plot as Figure 11.8, but displayed for the $N_z = 10$ zigzag systems under the distinct cases: (a) $(1)_s$, (b) $(9)_s$, (c) $(1,17)_d$, (d) $(1,9,17)_s$, (e) $(1,6,14,18)_d$, and (f) $(1,2,5,6,9,10,13,14,17,18)_s$.

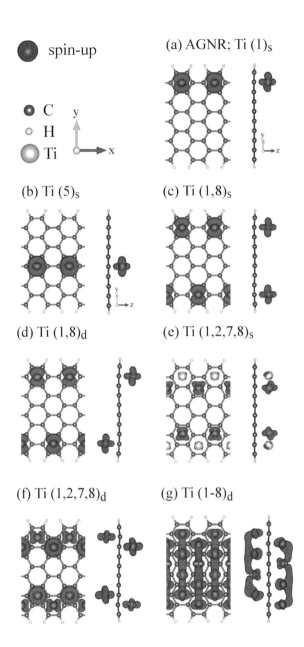

**FIGURE 11.10**

The spatial spin distributions for the Ti-adsorbed $N_A = 10$ armchair systems arising from the various chemisorptions: (a) $(1)_s$, (b) $(5)_s$, (c) $(1,8)_s$, (d) $(1,8)_s$, (e) $(1,2,7,8)_s$, (f) $(1,2,7,8)_d$, and (g) $(1\text{-}8)_d$. They/their insets are shown on the $x - y/y - z$ plane.

and magnetic environments. If two/four Ti adatoms are very close to the distinct zigzag edges (Figures 11.11(d)–11.11(f)), the AFM configurations purely due to the adsorbates are created and the spin states of the edge carbons are destroyed. In general, the spin arrangement near the specific boundary is FM for most of Ti adsorptions. It changes into the AFM configuration for very high concentrations, e.g., $(1,2,5,6,9,10,13,14,17,18)_d$ (Figure 11.11(f)), where the middle adatoms do not induce the spin states. That is, the same edge or the distinct two edges display the AFM configurations simultaneously.

The transition metal adatoms have five kinds of $d$ orbitals, so that the multi-orbtial hybridizations will be clearly shown on the $x - z$, $y - z$ and $x - y$ planes. The chemical bonding between Ti and C is obvious and significant even for the single-adatom cases, regardless of the positions, such as $(1)_s$ and $(9)_s$ for $N_A = 10$ and $N_z = 10$, respectively (Figures 11.12(a) and 11.12(c)). Its strength grows with the increasing adatom concentration e.g., $(1 - 8)_d$ and $(1, 6, 14, 18)_d$ (Figures 11.12(c) and 11.12(d)). The $\pi$ bonding of carbon $2p_z$ orbitals is distorted after chemisorptions, while it is still extended along the longitudinal and transverse directions. The carbon-dominated conduction bands crossing the Fermi level also make certain important contributions to the high free carriers. For any absorption cases, the large charge transfers exist between Ti guest adatoms and carbon host atoms on the $x$-$z$ and $y$-$z$planes, as clearly revealed in Figures 11.12(e)–11.12(h). It can only identify the significant orbital hybridizations of $(3d_{xz}, 3d_{yz}, 3d_{xy}, 3d_{z^2}, 3d_{x^2-y^2})$ five orbitals and $2p_z$ orbital. As to the Ti-Ti chemical bondings, the charge distributions are easily observed only on the $y - z$ plane at high concentration of armchair systems, e.g., $(1 - 8)_d$ in Figure 11.12(b). The charge differences become obvious for any chemisorptions on the $x - z$ plane and/or the $y - z$ plane; furthermore, there are charge extensions and even overlaps on the $x - y$ plane at the optimal heights Ti adatoms (the insets in Figures 11.12(e)–11.12(h)). The observable five-orbital hybridizations in Ti-Ti bonds are responsible for the low-lying Ti-dominated energy bands, being also one of the critical factors in the creation of the conduction electron density/the metallic behavior. In addition, it is very difficult to examine in detail about which $3d$ orbitals will make most of contributions to orbital interactions in Ti-C bonds under the current numerical calculations.

The orbital- and spin-projected DOSs in Ti-adsorbed graphene nanoribbons, as clearly shown in Figures 11.13(a)–11.13(h), could provide the very complicated van Hove singularities and thus identify the significant multi-orbital hybridizations in Ti-C and Ti-Ti bonds. For any chemisorptions, there is an obvious DOS at the Fermi level, directly indicating the creation of the high free carrier density. This is closely related to carbon guest atoms and Ti quest adatom. The spin-split DOSs are revealed in most of absorption cases (Figures 11.13(a)–11.13(e)) except for the symmetric adatom distributions in zigzag systems (Figures 11.13(f)–11.13(h)). The results further illustrate the Ti-induced spin states and their strong competitions with edge-carbon magnetic arrangement by the Ti-C chemical bondings. As to various orbital con-

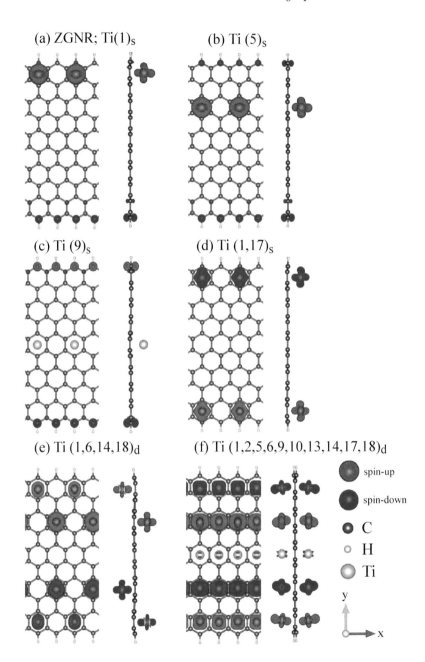

**FIGURE 11.11**
Similar plot as Figure 11.4, but shown for the $N_z = 10$ zigzag systems under the different chemisorptions: (a) $(1)_s$, (b) $(5)_s$, (c) $(9)_s$, (d) $(1,17)_s$, (e) $(1,6,14,18)_d$, and (f) $(1,2,5,6,9,10,13,14,17,18)_d$.

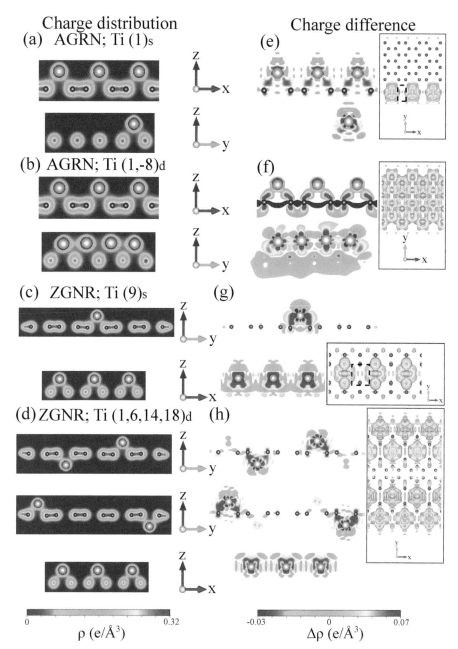

**FIGURE 11.12**
The spatial charge distributions and the differences after chemisorptions for the Ti-adsorbed systems. $\rho/\Delta\rho$ on the $x$-$z$ plane for $N_A = 10$ with Ti adatoms at (a)/(e) $(1)_s$ and (b)/(f) $(1\text{-}8)_d$. Also shown in the insets are $\Delta\rho$'s on the $x-y$ plane. Similar plots displayed for $N_z = 10$ under Ti-adatom positions: (c)/(g) $(9)_s$ and (d)/(h) $(1,6,14,18)_d$.

tributions to DOSs, $2s$, $2p_x$ and $2p_y$ of carbons (purple, blue and green curves) appear at $E \leq -3.4$ eV even for the highest atadom concentration (($1$-$8$)$_d$ in Figure 11.13(d)). Such evidence strongly suggests their interactions with the outer five $3d$ orbitals of Ti adatoms should be weak. However, the $2p_z$ orbitals experience rather high hybridizations with the $3d_{xy}$, $3d_{xz}$, $3d_{yz}$, $3d_{z^2}$ and $3d_{x^2-y^2}$, since the special structures of DOSs are seriously merged together in the range of $-1$ eV$\leq E \leq 3$ eV/$-2$ eV$\leq E \leq 3$ eV for the dilute/heavy adsorptions (Figures 11.13(a)–11.13(b) and Figures 11.13(e)–11.13(h)/Figures 11.13(c) and 11.13(d)). The Ti-Ti bonding is displayed in the enhancement of the metallic $3d$-band width. For example, energy band widths in the single-adatom and highest concentrations are, respectively, $\sim$1.5 eV and 3.0 eV (Figures 11.13(a) and 11.13(d)).

## 11.3   Bi

Bismuth adatom chemisorption on graphene nanoribbon surface can create the unusual geometric structure, being sensitive to the position and concentration. A single Bi adatom near the ribbon center covers $(5)_s$ and $(9)_s$ adsorptions in the $N_A = 10$ and $N_z = 10$ systems, respectively, as shown in Figures 11.14(a) and 11.14(b). The optimal position is located at the bridge site and its height is about 3.85 Å (Table 11.3), as abserved in 2D Bi-adsorbed graphene. When this adatom is close to the armchair/zigzag edge (Figures 11.14(c) and 11.14(e)), it shifts to the hollow site (along the $\hat{y}$ direction) about 0.3-0.4 Å (Table 11.3). Apparently, the height is reduced to the range of 2.4-2.5 Å. The similar results are foumd in the two-atom edge cases, e.g., the Bi heights of $(1,10)_s$ and $(2,9)_s$ adsorptions (Figure 11.14(c)). It is very difficult to form the high-concentration absorption system, since the Bi adatoms are very large and they have the significantly repulsive interactions within the sufficiently short distance. The maximum concentration is numerically examined to be about four and six Bi adatoms in the $N_A = 10$ and $N_z = 10$ systems, respectively (Figures 11.14(d) and 11.14(f)). The former even leads to the nonplanar/wave-like armchair graphene nanoribbon, indicating the quite strong Bi-C chemical bondings. Furthermore, there exist the abnormal heights in the peak and trough positions (Table 11.3), as shown by the $y - z$ plane projectionins for the $(1,4,7,10)_d$ adsorption of $N_A = 10$ (inset in Figure 11.14(d)). It is relatively difficult to observe the buckling structure in zigzag systems because of the larger widths. Most of the optimal structures show the planar geometries, or few of them display the wave-like ones accompanied with the bridge sites. As a result, the $\sigma$ bondngs of carbon atoms are hardly affected by Bi chemisorptions. The above-mentioned significant features clearly indicate the competition/cooperation among Bi-C, C-C and C-H bonds.

All the Bi-adsorbed graphene nonoribbons are 1D metals with free car-

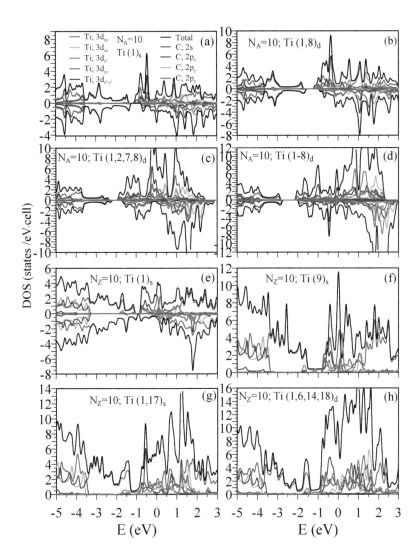

**FIGURE 11.13**

The orbital- and spin-decomposed DOSs of the $N_A = 10$ Ti-adsorbed systems under the various chemisorptions: (a) $(1)_s$, (b) $(1,8)_d$, (c) $(1,2,7,8)_d$, (d) $(1,2,3,4,5,6,7,8)_d$, and those for the $N_z = 10$ ones with the adatom positions: (e) $(1)_s$, (f) $(9)_s$, (g) $(1,17)_d$, and (h) $(1,6,14,18)_d$.

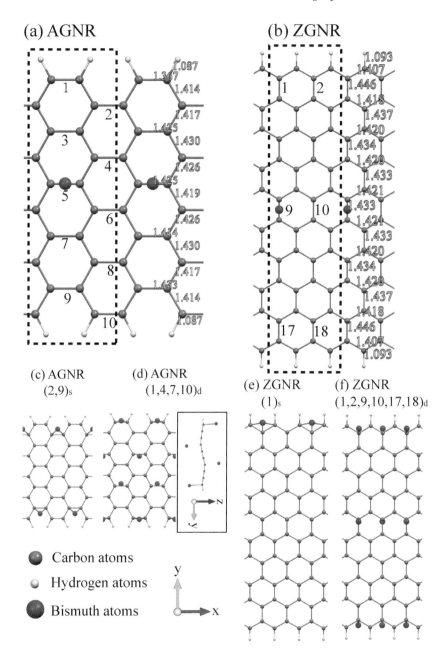

**FIGURE 11.14**

The optimal geometric structures of the Bi-adsorbed $N_A = 10$ armchair graphene nanoribbons with the initial adatom positions: (a) $(5)_s$, (c) $(2,9)_s$, and (d) $(1,4,7,10)_d$; those of the $N_Z = 10$ systems with the similar positions: (b) $(9)_s$, (e) $(1)_s$, and (f) $(1,2,9,10,17,18)_d$.

**TABLE 11.3**
Binding energy, adatom height, adatom y-shift, magnetic moment/magnetism for Bi-adsorbed $N_A = 10$ armchair and $N_Z = 10$ zigzag GNRs under various distributions and concentrations.

| | Eb (eV) | Height (Å) | Adatom y-shift (Å) | M ($\mu$B) |
|---|---|---|---|---|
| AGNR;Bi(1)$_s$ | -0.786 | 2.397 | 0.502 | 0.56/FM |
| (2)$_s$ | -0.071 | 2.483 | 0.551 | 0.11/FM |
| (5)$_s$ | -0.033 | 3.854 | x | 1.18/FM |
| (1,10)$_s$ | -0.443 | 2.404 | 0.481 | 1.08/FM |
| (2,9)$_s$ | -0.024 | 2.442 | 0.571 | 0.21/FM |
| (1,2,9,10)$_d$ | -0.615 | 1.933/3.319 | 1.011; -0.529 | 1.35/FM |
| (1,4,7,10)$_d$ | -0.276 | 2.521/3.046 | -0.235; 0.154 | 0.21/FM |
| ZGNR;Bi(1)$_s$ | -1.142 | 2.325 | 1.648 | 0.54/FM |
| (5)$_s$ | -0.697 | 3.944 | x | 1.59/FM |
| (9)$_s$ | -0.319 | 3.889 | x | 0/AFM |
| (1,17)$_s$ | -1.233 | 2.260 | 1.653 | 0/AFM |
| (1,9,17)$_s$ | -0.909 | 2.294/2.247 | 1.536/x | 0/AFM |

riers arising from the adatom chemisorptions, as clearly revealed in Figures 11.15 and 11.16. The adatom-dominated energy bands occur near the Fermi level, in which they belong to the low-lying valence and conduction bands. Roughly speaking, they come to exist in the range of the original energy gap, especially for the central and dilute adsorptions. Such bands will have wider energy widths in the increase of adatom concentration. Most of them might have the weak energy dispersions with the high DOSs. The $\sigma$ bands ($\sim E^v \leq -3$) due to carbon atoms are almost independent of Bi adtoms except for energy shifts, clearly reflecting the optimal geometric structures. However, the significant (C,Bi)-co-dominated band structure could occur at low energy. Moreover, whether band structures exhibit the spin-split behaviors depends the edge structure and the adatom distribution.

All the armchair Bi-adsorbed systems possess the spin-split electronic states, especially for the adatom-dominated energy bands. In general, the spin splittings are not obvious in the carbon-dominated energy bands. The occupied states near the Fermi level are not equivalent in the spin-up and spin-down ones, clearly indicating the net magnetic moment in each system. This is in great sharp with the non-magnetic behavior of the Al chemisorptions (the non-spin-split energy bands in Figure 11.2). The magnetic properties are almost fully determined by the Bi adatoms in the armchair graphene nanoribbons (Figure 11.15; discussed in Figure 11.17). The net magnetic momentum is very sensitive to the adatom configuration (Table 11.3), and the maximum

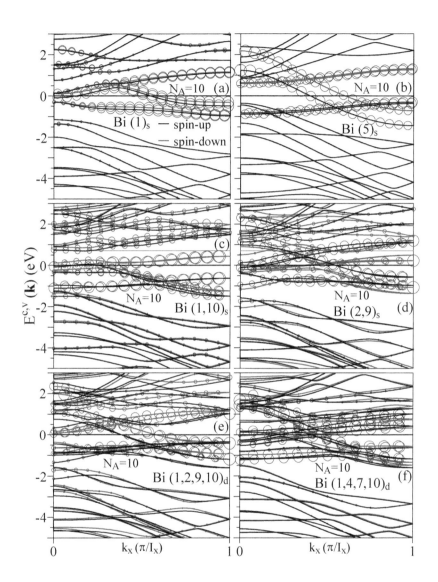

**FIGURE 11.15**

Energy bands for Bi-adsorbed $N_A = 10$ armchair graphene nanoribbons under the distinct adsorptions: (a) $(1)_s$, (b) $(5)_s$, (c) $(1,10)_s$, (d) $(2,9)_d$, (e) $(1,2,9,10)_d$, and (f) $(1,4,7,10)_d$.

value could reach 1.35 $\mu_B$ per unit cell in the specific $(1,2,9,10)_d$ adsorption. As to the 1D conduction electrons, they are closely related to the Bi-dependent energy bands crossing the Fermi level. Compared with one free electron per adatom in alkali chemisorptions (Section 9.1), Bi adatom could induce more free carriers only under the single-adatom chemisorption, while the opposite is true for other adsorption cases.

The low-energy electronic structures are greatly diversified by the edge structure, and the zigzag and armchair Bi-adsorbed systems are very different from each other (Figures 11.16 and 11.15). The edge-carbon- and Bi-induced energy bands in the zigzag nanoribbons appear near the Fermi level simultaneously, if the adatom is situated near the ribbon center (e.g., the $(9)_s$ case in Figure 11.16(c)). The former remain a pair of partially flat dispersions crossing the Fermi level (without blue circles), without the spin splitting. Furthermore, the latter exhibits the unusual behavior in the absence of spin-split electronic states; that is, it does not have any magnetism. This unique magnetic property will be thoroughly discussed in the spatial spin distribution (Figure 11.18(c)). The low-lying energy bands will become very complicated for the adatoms close to the zigzag boundaries, such as the $(1)_s$ and $(5)_s$ adsorptions in Figures 11.16(a) and 11.16(b), respectively. Specifically, the $(1)_s$ case exhibits the drastic changes in the originally edge-localized bands, and it has the strongly hybridized energy bands in the range of $-1$ eV$\leq E^{c,v} \leq 1$ eV in the presence of significant contributions from edge carbons and bismuths simultaneously. These clearly indicate the important chemical bondings of host and quest atoms. The similar results could be found in the higher-concentration edge adsorptions, e.g., the $(1,17)_s$, $(1,9,17)_s$ and $(1,2,9,10,17,18)_d$ chemisorptions in Figures 11.16(d), 11.16(e), and 11.16(f), respectively. As to the magnetic properties, most of the Bi-adsorbed zigzag nanoribbons display the spin splittings and the FM configuration (Table 11.3; spin density in Figure 11.18). It is very easy to measure the magnetic response, since the significant magnetic moment is about $0.5 - 1.6$ $\mu_B$ per unit cell. Only few of absorption configurations possess the spin-degenerate energy bands: one-adatom center (Figure 11.16(c)), two-adatom zigzag edges (Figure 11.16(d)), and three-adatom center and edges (Figure 11.16(e)). Apparently, the net moment is vanishing and the spin arrangement belongs to the AFM configuration.

The spin configuration (Figures 11.17 and 11.18) and the strength of magnetic response (Table 11.3) are greatly diversified by the Bi chemisorptions on the graphene nanoribbon surfaces. The isolated bismuth atoms could themselves induce the intrinsic magnetic mement, so only the FM and AFM spin distributions come to exist after the quest-adatom adsorptions. All the armchair systems show the FM configurations, and most/few of the zigzag ones exhibit the FM/AFM spin arrangement. For the former, the Bi guest atoms create the spin-up distribution, also leading to the minor and similar magnetic configuration of the nieghboring carbon atoms (Figures 11.17(a)–11.17(g)). From the single-adatom cases, the spatial spin densities are strongly dependent on its position, e.g., the obvious differences among the $(1)_s$, $(2)_s$ and $(5)_s$

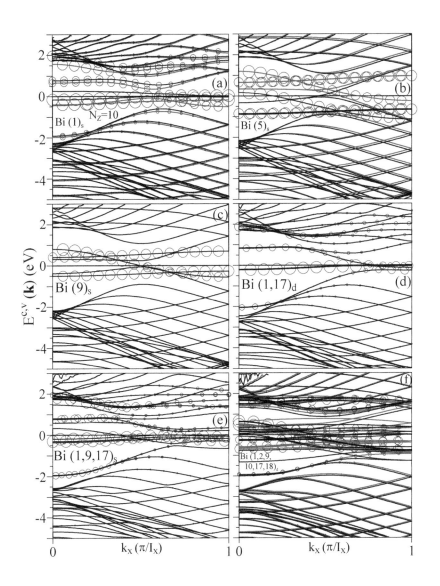

**FIGURE 11.16**

Similar plot as Figure 11.15, but shown for the $N_Z = 10$ zigzag systems under the various cases: (a) $(1)_s$, (b) $(5)_s$, (c) $(9)_s$, (d) $(1,17)_s$, (e) $(1,9,17)_s$, and (f) $(1,2,9,10,17,18)_d$.

chemisorptions (Figures 11.17(a)–11.17(c); 0.56 $\mu_B$, 0.11 $\mu_B$ and 1.18 $\mu_B$ in Table 11.3). This is related to the chemical environment experienced by the Bi adatoms. The magnetic responses are also very sensitive to the adatom concentration, in which, a simple linear relation between them is absent. For example, the net magnetic moment is large under the $(1, 10)_s$ and $(1, 2, 9, 10)_d$ adsorptions (Figures 11.17(d) and 11.17(f); 1.05 $\mu_B$ and 1.35 $\mu_B$), while it is small for the $(2, 9)_s$ and $(1, 4, 7, 10)_d$ adsorptions (Figures 11.17(e) and 11.17(g); 0.21 $\mu_B$). The magnetic properties are also affected by the terminated C-H bonds, as implied from two-adatom cases. Four-adatom chemisorptions show that the neighboring Bi adatoms could enhance the strong magnetic response. However, the edge and near-center adatoms, respectively, create the spin-down and spin-up distributions and thus induce the weaker magnetic response.

On the other side, the zigzag systems have the quite strong competition/cooperation between Bi adatoms and edge carbons in the spin state/orbital hybridization, leading to the diverse and unique magnetic phenomena. When the Bi-adsorbed system remains a symmetric geometry under the single adsorptions, the adatom simultaneously exhibits the spin-up and spin-down magnetic moments with the same magnitude ($(9)_s$ in Figure 11.18(c)), or the Bi-dependent magnetism is thoroughly vanishing. Also, the spin arrangement of edge carbons keeps the AFM configuration across the ribbon center. However, the Bi adatom in an asymmetric edge distribution itself only generates the spin-up state, e.g., $(1)_s$ and $(5)_s$ (Figures 11.8(a) and 11.8(b)). It has a strong effect on the edge-carbon spins and even alters the magnetic dominance in the zigzag edges. The obvious FM configurations are revealed along the longitudinal and transverse directions. Specifically, the $(5)_s$ adsorption shows the maximum magnetic moment ($1.59\mu_B$ in Table 11.3).

The group-V elements have three valence electrons in the outmost orbitals, they can form the complex multi-orbital hybridizations after chemisorptions on graphene nanoribbon surfaces. In general, the Bi adatoms possess the $(6p_x, 6p_y, 6p_z)$ orbitals which will take part in the Bi-C and Bi-Bi bonds. The multi-orbital hybridizations are clearly revealed in the spatial charge distributions, as observed in Figure 11.19. The single-adatom central adsorptions are shown for armchair and zigzag graphene nanoribbons in Figures 11.19(a)/11.19(d) and Figures 11.20(a)/11.20(d), respectively, in which $\rho$ and $\Delta\rho$ almost have no charge distributions between guest and host atoms. This suggests the rather weak interactions in Bi-C bonds under the very high optimal position ($3.85 - 3.89$ Å in Table 11.3). However, Figures 11.19(d) and 11.20(d) in $x - z$ and $x - y$ planes clearly show that charge variation is obvious near the Bi adatoms, indicating the significant Bi-Bi bonds. The main orbital hybridizations cover $6p_x - 6p_x$, $6p_y - 6p_y$ and $6p_z - 6p_z$, e.g., the first one in the armchair system (the black dashed rectangle in the inset of Figure 11.19(d)). On the other hand, for most of Bi-adatom adsorptions (the others), there exist the observable/apparent charge distributions in Bi-C bonds and more complicated orbital interactions in Bi-Bi bonds. That is, the asymmetric single-adatom and higher-concentrations adsorptions could create more

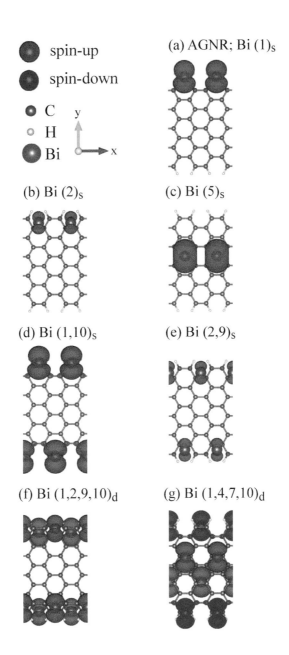

**FIGURE 11.17**
The spatial spin distributions of the Bi-adsorbed $N_A = 10$ armchair systems
due to the various adsorptions: (a) $(1)_s$, (b) $(2)_s$, (c) $(5)_s$, (d) $(1,10)_s$, (e)
$(2,9)_d$, (f) $(1,2,9,10)_d$, and (g) $(1,4,7,10)_d$. They are shown on the $x - y$ plane.

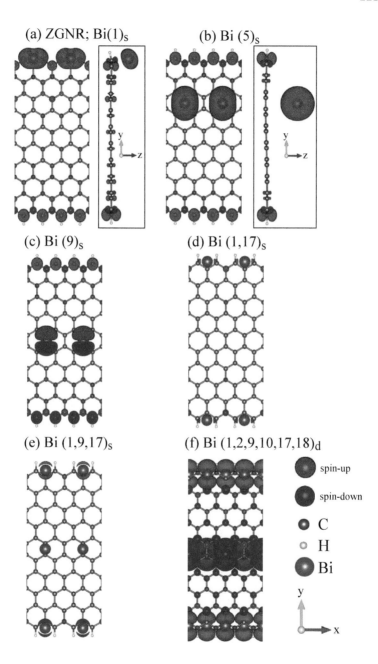

**FIGURE 11.18**
Similar plot as Figure 11.17, but displayed for the $N_Z = 10$ zigzag systems for the different cases: (a) $(1)_s$, (b) $(5)_s$, (c) $(9)_s$, (d) $(1,17)_s$, (e) $(1,9,17)_s$, and (f) $(1,2,9,10,17,18)_d$. Also shown in the insets of (a) and (b) are those on the $y - z$ plane.

orbital hybridizations in Bi-C and Bi-Bi bonds, such as, $(1)_s$ and $(1,4,7,10)_d$ in armchair systems (Figures 11.19(b) and 11.19(c)); $(1)_s$ and $(1,9,17)_d$ in zigzag cases (Figure 11.20(b) and 11.20(c)). The significant Bi-C bondings have induced the charge distributions between them, accompanied with the partial distortions of the $\pi$ bondings. The distorted and extended $\pi$ bondings and the Bi-Bi bonds dominate the metallic behavior, especially for the latter. This is consistent with the orbital-projected DOSs (discussed in Figure 11.19). The important orbital hybridizations in Bi-C bonds, as observed from the charge differences, cover $6p_z - 2p_z$ (clearly indicated in Figure 11.19(e)/Figure 11.19(f)/Figure 11.20(e)/Figure 11.20(f)), $6p_z - 2p_y$ (Figure 11.20(e)/Figure 11.20(f)), and $6p_z - 2p_x$ (Figure 11.19(e)/Figure 11.19(f)).The first kind has the strongest orbital hybridization among them.

The orbital- and spin-decomposed DOSs obviously reveal the metallic behavior, the magnetic configuration, and the concise orbital hybridizations in chemical bonds. The Bi adatom chemisorptions can induce a finite DOS at the Fermi level (Figures 11.21(a)–11.21(h)), in which its value directly reflects the energy dispersions crossing $E_F = 0$ (Figures 11.15 and 11.16). Most of cases exhibit very high DOSs (Figures 11.21(c)–11.21(h)), accompanied with the weakly dispersive bands. However, the dilute absorptions in armchair systems might display low DOSs ($(1)_s$ and $(5)_s$, respectively, in Figures 11.21(a) and 11.21(b)) because of the highly dispersive bands. The spin-up and spin-down DOSs are different in all the Bi-adsorbed graphene nanoribbons except for the symmetric adatom distributions in zigzag systems (Figures 11.21(f)–11.21(h)). Their difference in the covered area means the FM configuration and is proportional to the net magnetic moment. For example, it is about half for the ratio between the $(1)_s$ and $(5)_s$ armchair cases, and so does the net magnetic moment (0.56 $\mu_B$ and 1.18 $\mu_B$ in Table 11.3). The AFM and NM configurations cannot be distinguished from the spin-degenerate DOS, while they are confirmed by the spatial spin distributions (Figures 11.18(c)–11.18(e)). As to the orbital hybridizations, the DOSs, which arise from the $(2s, 2p_x, 2p_y)$ orbitals of carbon atoms, occur at $E \leq 3$ eV and are hardly affected by the surface adsorptions. That is, the planar $\sigma$ bonding almost keeps the similar, being consistent with the optimal geometric structures. Only the $2p_z$ orbitals might take part in the Bi-C bonding. In general, their contributions to the total DOS in the range of $-1.5$ eV$\leq E \leq 2.5$ eV show the special structures merged those arising from the Bi $(6p_x, 6p_y, 6p_z)$ orbitals. The $6p_z - 2p_z$ bondings (the orange and red curves) are strongest among three kinds of orbital hybridizations in Bi-C bonds, as clearly indicated from the strength and number of combined van Hove singularities. However, the emergence is greatly reduced in the single-adatom central adsorptions $(5)_s$ and $(9)_s$, respectively, in armchair and zigzag systems. This is responsible for the very weak Bi-C bonding. As to the significant hybridizations of $(6p_x, 6p_y, 6p_z)$ in Bi-Bi bandings, they are obviously revealed in the three-orbital-dependent merged structures, in which the intensities are enhanced in the increase of adatom concentration. The $6p_x$-$6p_x$ and $6p_y$-$6p_y$ orbital hybridizations are

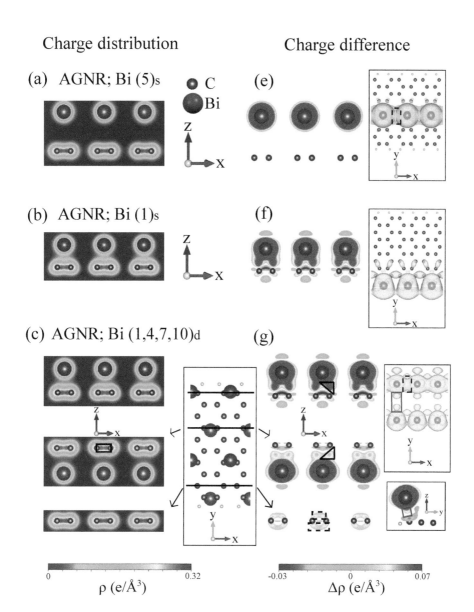

## FIGURE 11.19

The spatial charge distributions and the differences after chemisorptions. $\rho$'s on the $x-z$ plane for the $N_A = 10$ systems with Bi adatoms at (a) $(5)_s$, (b) $(1)_s$, and $(1,4,7,10)_d$, respectively, correspond to $\Delta \rho$'s in (d), (e), and (f). The latter are also indicated on the $x-y$ plane in the insets.

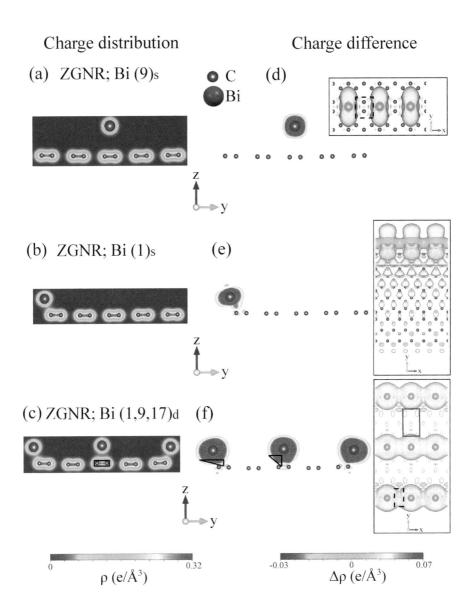

**FIGURE 11.20**
Same plot as Figure 11.19, but shown for the $N_Z = 10$ systems under $\rho$'s on the $y - z$ plane.

**TABLE 11.4**
Binding energy, adatom height, adatom y-shift, magnetic moment/magnetism for Fe-, Co-, Ni-adsorbed $N_A = 10$ armchair and $N_Z = 10$ zigzag GNRs under various distributions and concentrations.

| | Eb (eV) | Height (Å) | Adatom y-shift (Å) | M ($\mu$B) |
|---|---|---|---|---|
| AGNR; Fe $(1)_s$ | -1.125 | 1.606 | 0.018 | 2.37/FM |
| $(5)_s$ | -0.848 | 1.643 | 0.002 | 2.40/FM |
| $(1,8)_s$ | -2.835 | 1.609 | 0.013 | 4.74/FM |
| $(1,8)_d$ | -2.835 | 1.639 | 0.006 | 4.74/FM |
| $(2,7)_s$ | -2.528 | 1.608 | 0.007 | 4.63/FM |
| $(2,7)_d$ | -2.529 | 1.596 | 0.003 | 4.78/FM |
| $(1,2,7,8)_d$ | -3.521 | 1.605/1.591 | 0.055;0.047 | 9.86/FM |
| $(1,2,3,4,5,6,7,8)_d$ | -4.723 | 1.932/1.973 | 1.019;1.037 | 21.85/FM |
| ZGNR; Fe $(1)_s$ | -1.227 | 1.659 | 0.005 | 2.40/FM |
| $(5)_s$ | -0.246 | 1.527 | 0.006 | 0.03/FM |
| $(9)_s$ | -0.182 | 1.471 | x | 0/AFM |
| $(1,17)_s$ | -2.840 | 1.658 | 0.001 | 0/AFM |
| $(1,9,17)_s$ | -3.055 | 1.658 | 0.008;x | 0/AFM |
| $(1,6,14,18)_d$ | -3.454 | 1.669 | 0.007;0.004 | 0/AFM |
| $(1,2,5,6,9,10,13,14,17,18)_d$ | -3.857 | 1.675 | 0.008;0.002;x | 0/AFM |

significant and dominating, as observed from the orbital-dependent van Hove singularities.

## 11.4 Fe/Co/Ni

The Fe/Co/Ni-adatom chemisorptions on 1D graphene nanoribbons could create the optimal geometric structures similar to 2D graphene systems except for the non-uniform band lengths. The Fe heights are in the range of $\sim 1.40 - 2.00$. Their binding energies are larger under higher concentrations, as shown in Table 11.4. It is relatively easy to induce the high concentrate Fe chemisorptions, compared with the above-mentioned metal adatoms. The Fe adatoms could reach/present the higher adatom concentrations as well as Ti adatoms. Specifically, the buckling structures, might be revealed for the highest-concentration adsorption in armcahir/zigzag systems, which similar to the Ti absorptions.

Any Fe/Co/Ni-adsorbed graphene nanoribbons belong to the unusual metallic systems, as shown in Figures 11.22 and 11.23. There are several en-

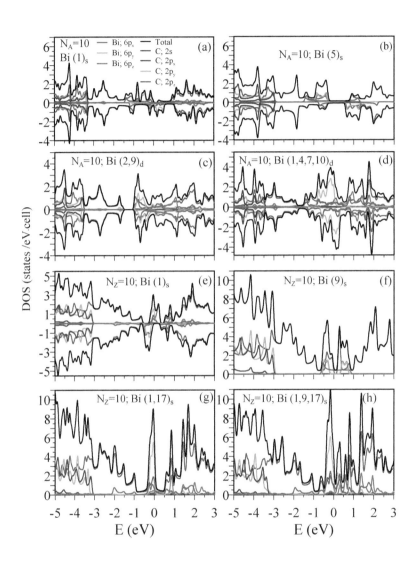

**FIGURE 11.21**
The orbital- and spin-decomposed DOSs for the $N_A = 10$ armchair systems under the various cases: (a) $(1)_s$, (b) $(5)_s$, (c) $(2,9)_d$, (d) $(1,4,7,10)_d$, and those for the $N_Z = 10$ systems with the adatom adsorptions: (a) $(1)_s$, (b) $(9)_s$, (c) $(1,17)_s$, and (d) $(1,9,17)_s$.

ergy bands crossing the Fermi level, in which they mainly come from both Fe adatoms and carbon atoms. Furthermore, the Fe-dominated energy bands might be thoroughly occupied or unoccupied. For armchair systems, all of them exhibit the spin-split energy bands, especially for those near $E_F$. They correspond to the FM spin configurations, and the net magnetic moments due to the Fe adsorbates are sensitive to the adatom distribution and concentration (Table 11.4). Under the single- and two-adatom adsorptions (Figures 11.22(a)–11.22(d)), the Fe-dependent electronic structures lie in the range of $-3$ eV$\leq E^{c,v} \leq 2$ eV, and the C atoms fully determine the energy bands below it. The increasing adatom concentration, respectively, makes major and minor contributions to these two different ranges of energy bands. The Fe adatoms hardly contribute to the deep valence states, indicating the weak chemical hybridizations from the $(2s, 2p_x, 2p_y)$ orbitals in the Fe-C bonds. As for Co adatoms, they dominate the low-lying energy bands as well as Fe adatoms do, presenting similar spin-split spacing at low energy. However, The Ni adatoms possess no magnetic moments, resulting in the absence of band split in low energy. It should be noticed that the Ni adatoms also contribute to the energy band in the low energy range.

The Fe/Co/Ni chemisorptions could induce the diversified band structures by the edge structures. There exist important differences between the zigzag and armchair systems. As for the Fe-adsorbed ones, the partially flat bands across the Fermi level only survive under the single-adatom central adsorption (Fe systems in Figure 11.23(b)). However, most of cases, being shown in Figures 11.23(a) and 11.23(c) and 11.23(d), revealing the dramatic changes in the low-lying energy bands because of the significant Fe-edge-C bondings; that is, the edge-C-dominated partially flat energy dispersions are thoroughly absent. For the Co-adsorbed systems, the low-lying energy bands are similar to Fe systems, as shown in Figure 11.23(e). However, Ni-adsorbed ones exhibit smaller spin split in low-lying bands, which owing to the weak magnetic moments.

The orbital- and spin-projected DOSs in Fe/Co/Ni-adsorbed graphene nanoribbons, as clearly shown in Figures 11.24(a)–11.24(h), could provide the very complicated van Hove singularities and thus identify the significant multi-orbital hybridizations in Fe/Co/Ni-C and bonds between themselves. For all systems, there is an significant DOS near the Fermi level, directly indicating the creation of the high free carrier density. This is closely related to carbon guest atoms and Fe/Co/Ni quest adatom. The spin-split DOSs are revealed in most of absorption cases (Figures 11.24(a)–11.24(c); 11.24(e); 11.24(g); 11.24(h)) except for the symmetric adatom distributions in zigzag systems (Figure 11.24(f)) and the Ni-adsorbed armchair nanoribbons (Figure 11.24(d)). These results indicate the Fe-induced spin states and their competitions with edge-C in magnetic arrangement by the Fe-C chemical bondings. As for various orbital contributions to DOSs, $2s$, $2p_x$ and $2p_y$ of carbons (purple, blue and green curves) appear at $E \leq -3.4$ eV even for the different atadom concentration $((1,8)_s$ in Figure 11.24(b)). This illustrates that their interac-

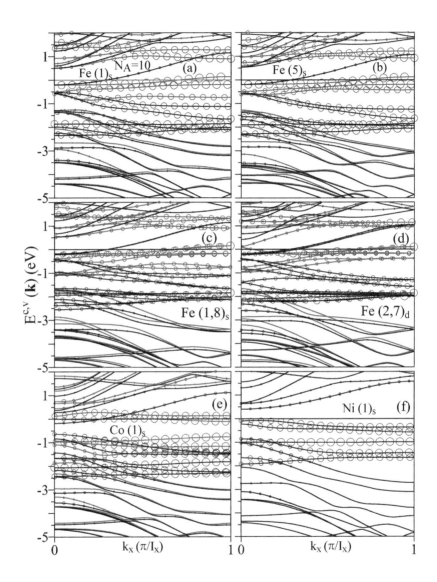

**FIGURE 11.22**
Energy bands for $N_A = 10$ armchair graphene nanoribbons under the distinct adsorptions: (a) Fe(1)$_s$, (b) Fe(5)$_s$, (c) Fe(1,8)$_s$, (d) Fe(2,7)$_d$, (e) Co(1)$_s$, and (f) Ni(1)$_s$.

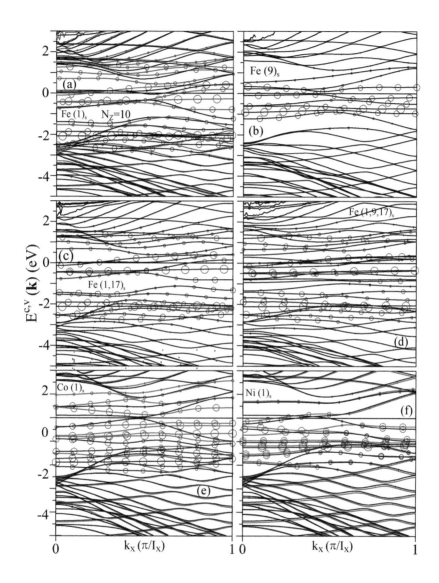

**FIGURE 11.23**
Energy bands for $N_A = 10$ zigzag graphene nanoribbons under the distinct adsorptions: (a) Fe(1)$_s$, (b) Fe(9)$_s$, (c) Fe(1,17)$_s$, (d) Fe(1,9,17)$_s$, (e) Co(1)$_s$, and (f) Ni(1)$_s$.

tions with the outer five $3d$ orbitals of Fe adatoms should be weak. Further-more, the $2p_z$ orbitals exhibit hybridizations with the $3d_{xy}$, $3d_{xz}$, $3d_{yz}$, $3d_{z^2}$ and $3d_{x^2-y^2}$, which revealed by the special structures of DOSs in the range of $-1$ eV$\leq E \leq 3$ eV. The Co-adsorbed systems present similar structures in both armchair and zigzag nanoribbons, including the van Hove singularities, the DOS near the Fermi level, and the hybridizations between $2p_z$ and $3d$ orbitals (Figures 11.24(c) and 11.24(g)). However, the Ni-adsorbed armchair ribbons has no spin-split states (Figure 11.24(d)), and the correspond zigzag systems possess rather symmetric peaks on spin-up and spin-down states (Figure 11.24(h)).

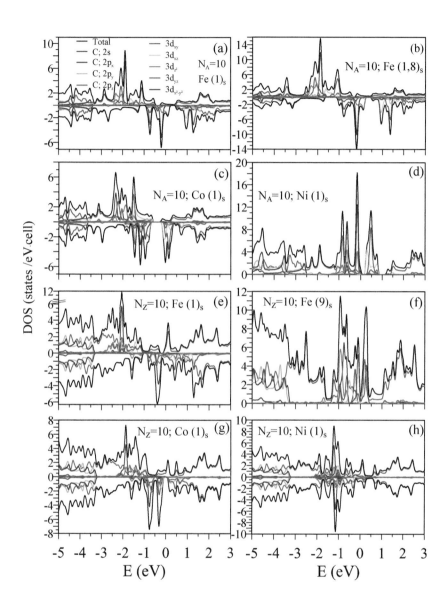

**FIGURE 11.24**
The orbital- and spin-decomposed DOSs for the $N_A = 10$ armchair systems
under the various cases: (a) Fe$(1)_s$, (b) Fe$(1,8)_s$, (c) Co$(1)_s$, (d) Ni$(1)_s$, and
those for the $N_Z = 10$ systems with the adatom adsorptions: (a) Fe$(1)_s$, (b)
Fe$(9)_s$, (c) Co$(1)_s$, and (d) Ni$(1)_s$.

# 12

## Concluding remarks

This systematic work is conducted on the geometric strictures, electronic properties and magnetic configurations of the 1D graphene nanoribbons. They are thoroughly investigated by the first-principles calculations. The rich and unique essential properties arise from the various geometries, adatom dopings and spin arrangements. There exist the planar, folded, curved, zipped, stacked, edge-passivated, and adatom-adsorbed nanoribbon structures. The responsible mechanisms covers the $sp^2$ and $sp^3$ bondings in the angle-dependent C-C bonds, the significant orbital hybridizations in adatom-adatom bonds, and the single- or multi-orbital hybridizations in C-adatom bonds. Electronic structures exhibit the semiconducting and metallic behaviors, accompanied with the non-magnetism, anti-ferromagnetism and ferromagnetism. A lot of factors, the hexagonal/non-hexagonal lattice, non-uniform bond lengths, finite-size confinement, edge structure, non-planar structure, stacking configuration, layer number, boundary condition, distribution and concentration of adatoms, Van der Waals interactions, and spin-dependent many-body effects, are taken into consideration to get the concise physical and chemical pictures. These will be very helpful in investigating materials science and applications The calculated results are compared with those from the previous theoretical predictions. Part of them have been confirmed by the experimental measurements. Specially, a theoretical framework is developed by means of the first-principles calculations. The critical orbital hybridizations in various chemical bonds, which diversify the essential properties, are obtained from the detailed analyses on the atom dominance of energy bands, the spatial charge distributions & density differences, and the orbital-projected DOSs. Furthermore, the edge-carbon- and adatom-dependent magnetic configurations are identified from the net magnetic moment, the spin-split energy bands, the spin density distributions, and the spin-decomposed DOSs. This method could be further improved and generalized to explore the chemically modified materials.

Graphene nanoribbons exhibit the rich essential properties, being very sensitive to the edge structure, width and hydrogen termination. The central C-C bond lengths are close to those of monolayer graphene, while the boundary ones become shorter in the presence and absence of edge dangling bonds, especially for the former case. In addition to the quantum confinement, the non-uniform hopping integrals (the AFM of edge carbons) could account for the semiconducting $N_A = 3I - 1$ armchair graphene nanoribbons (zigzag ones) [324]. As for armchair systems, energy gaps, which are inversely proportional to widths, could be classified into three kinds. They are, respectively, maximum and minimum for $N_A = 3I + 1$ and $N_A = 3I$ $(3I - 1)$ without (with) hydrogen passviations. The width-dependent rules are also exist in direct-gap zigzag graphene nanoribbons. All the graphene nanoribbons are direct-gap semiconductors except for $N_A = 3I$ armchair systems. Specifically, energy gaps of zigzag systems arise from the $k_x = 2/3$ states of the edge-localized valence and conduction bands. There are a lot of parabolic bands and some partially flat bands, in which the latter come from edge carbons. They, respectively, show the square-root asymmetric peaks and the delta-function-like symmetric ones in DOSs. The band-edge states, creating the van Hove singularities, occur at $k_x = 0, 1$ and in between. All (most) of energy bands are due to the $\pi$ and $\sigma$ bondings under the edge passivations (the free boundaries). The slight $sp^3$ bondings only appear in the dangling bonds, revealed by the merged conduction-band peaks of $(2s, 2p_x, 2p_y, 2p_z)$ orbitals in DOSs and the partially flat valence/conduction band in armchair/zigzag graphene nanoribbons at about $-2.5 - 2.2$ eV$/0.9 - 1.0$ eV. There exist the $sp^2 s$ hybridyizations in the C-H bonds, since the $(2s, 2p_x, 2p_y, 1s)$-decomposed DOSs occur simultaneously. Due to the finite-size effect, the $2p_x$- and $2p_y$-decomposed DOSs might be quite different from each other in terms of peak position and intensity. The low-lying strong peak structures lies in the $2p_z$ orbitals. A pair of peaks, which is centered at the Fermi level, can characterize energy gap. Only carbons in zigzag edges can induce spin configurations, AFM across nanoribbon center and FM on the same side. The spin-polarized ground state energy is lower than non-spin-polarized one, being sensitive to the nanoribbon width.

The curvature, width/radius, edge structure/chiral angle, and boundary condition can create the diverse essential properties in curved graphene nanoribbons and carbon nanotubes. During the zipping process of a graphene nanoribbon, the edge-atom bond length and the ground state energy dramatically vary with the arc angle. The critical zipping and unzipping energies, arc angle, and interaction distance are sensitive to the width and edge structure. They depend on the very strong competition between the mechanical strain and the chemical bonding of the two-side edge atoms. Apparently, the increasing curvatures induce the drastic changes of electronic properties, such as energy gaps, weak or strong energy dispersions, band-edge states, band mixing, band overlap, state degeneracy, complex orbital hybridizations, and van Hove singularities in DOS. The semiconductor-metal transitions occur near the critical arc angle, directly reflecting the contribution of the edge atoms, the competition between $\pi$ and $\sigma$ bondings, and the hybridization of the $2p_z$ and $2p_y$ orbitals. The predicted geometric structures and electronic properties could be examined by the STM, STS and ARPES measurements. As a result of the geometric symmetries and the boundary conditions, there exist important differences between armchair/zigzag nanotubes and zigzag/armchair nanoribbons in the main features of electronic properties. They cover the existence of the transverse quantum number, the contributions of the edge and non-edge carbon atoms to each energy band, the metallic or semiconducting behavior, and the anti-ferromagnetic or non-magnetic spin configuration. Three types of carbon nanotubes are: (I) metallic armchair and ($m \leq 6,0$) zigzag systems: (II) moderate-gap ($m = 3I \pm 1,0$) and (III) narrow-gap ($3I,0$) zigzag ones (inversely proportional to $r$ and $r^2$). The significant $sp^3$ orbital hybridization and the $2p_z$-orbital misorientation, respectively, lead to the metallic and narrow-gap zigzag nanotubes. All armchair nanotubes, with a specific hexagonal arrangement on a cylindrical surface, are 1D metals arising from the linearly intersecting valence and conduction bands at the Fermi level. However, the semiconducting zigzag nanoribbons present a pair of separated edge-localized energy bands because of the anti-ferromagnetic spin arrangement at the two-side zigzag edges. Both ($m > 6,0$) zigzag nanotubes and armchair nanoribbons exhibit the semiconducting property, in which a simple inverse relation between energy gap and radius/width appears for $m = 3I \pm 1$ and $N_A = 3I - 1, 3I, and 3I + 1$. Specifically, the smallest energy gaps in $N_A = 3I - 1$ graphene nanoribbons with hydrogen passivations are closely related to the non-uniform $\pi$ bondings at the distinct carbon positions.

The folded graphene nanoribbons, with the flat and curved structures, exhibits the rich geometric, electronic and magnetic properties, mainly owing to the combined confinement, edge, stacking and curvature effects. There are eight typical types of geometric structures, according to the achiral edge, the number of armchair/zigzag lines and the AA/AA'/AB stacking. The armchair AA' stacking or the even-zAB stacking is the most stable configuration. The edge-edge distance, the interlayer distance and the nanotube diameter are sensitive to the stacking configuration and ribbon width. Only the even-zAA stacking, with the mirror symmetry, exhibit a pair of metallic liner bands under the strong edge-edge interactions. The other stacking systems belong to the direct-gap semiconductors. In general, band gap declines with the increasing width because of the weakened finite-size effect. However, it almost keeps the same for any even-zAB stacking, being attributed to the breaking of mirror symmetry, and the saturated edge-edge distance and nanotube diameter. Six categories of width-dependent energy gaps are revealed in folded armchair nanoribbons, in a sharp contrast with three categories in planar systems. Specially, the $N_A + 6I = 4$ even-aAA' stacking has the highest energy gaps, compared with the various pristine graphene nanoribbons. The magnetic configurations depend on the edge structure, the edge-edge interaction, and the stacking symmetry. As to the main structures of DOS, they cover a plateau across $E_F$, strong peaks, and asymmetric peaks, respectively, arising from the linear, partially flat and parabolic bands. The non-magnetic behavior occur in all armchair systems and (even-zAA,even-zAA) stackings with the dominating edge-edge interactions. Both odd-zAA' and odd-zAB stackings present the anti-ferromagnetic spin distribution across two parallel open edges. For the latter, the different spin-up and spin-down environments lead to the spin splitting and decreasing energy gaps. The spin-polarized STS and ARPES are available in the identifications of the spin-dependent van Hove singularities and valence bands. Apparently, the folded structures have greatly diversified the lattice symmetries, the semiconducting and metallic properties, and the spin configurations, compared with the planar ones.

Carbon nanoscrolls, with the non-uniform curved surfaces, are formed under the strong competitions between the interalyer atomic interactions and the mechanical strains. Each optimal structure possesses a spiral profile after the self-consistent relaxations. The scrolling energy exhibits the declining and fluctuating behaviors in the increase of width because of the complex curvature variations. It is higher in armchair nanoscrolls, compared with zigzag systems. The latter are predicted to be relatively easily produced in experimental syntheses. The similar difference could also be observed in folded armchair and zigzag systems. To sustain a stable nanoscroll, it needs to have the sufficiently large ribbon width and inner diameter. Two critical geometric parameters depend on the edge structure. For armchair (zigzag) nanoscrolls, the minimum widths are $N_A =$34, 43; 47 ($N_Z =$18, 20; 22), respectively, corresponding to the critical diameters of $N_{in} = 7$, 9; 11 (4, 6; 8). The bilayer-like configuration is close/similar to the AB stacking, as revealed in the most stable bilayer graphene. Armchair systems have the significant orbital hybridizations related to the inner open edge and the curved surface, leading to the unusual low-lying energy dispersions and band gaps. These exist three categories of energy gaps according to $N_A = 3I, 3I + 1; 3I + 2$, in which their values are quite different from those of the planar armchair nanoribbons. On the other hand, both edge structure and quantum confinement also dominate the partially flat bands of zigzag systems. The spin-split edge-localized bands, with narrow energy gaps, come to exist when the spin-up and spin-down distributions experience the distinct magnetic environments. The various combined effects can induce the important differences among the planar, curved/zipped, folded and scrolled graphene nanoribbons, such as, the existence of the critical/saturated internal diameter and the stable stacking configurations, the categories of the width-dependent energy gaps, the parabolic/linear/oscillatory bands nearest to $E_F$, the semiconductor-metal transitions, and the breaking or conservation of spin degeneracy. This clearly illustrates that the boundary condition, the edge-dependent interactions, the single- or multi-orbital hybridizations and the magnetic configurations are responsible for the diversified essential properties.

The essential properties of bilayer zigzag graphene nanoribbons are greatly enriched by the relative shift between the upper and lower layers. They rely on the stacking configurations, interlayer edge-edge interactions, spin distributions, and ribbon widths. The AA and $AB_\alpha$ stackings exhibit the arc-shaped structure with the convex and wave-like profiles, respectively, in which the shorter edge-edge distances lead to the strong interactions and thus the destruction of magnetism. The most stable configuration in all the bilayer systems lies between them, mainly owing to the strong competition of the combined effects. The low-lying energy bands are composed of parabolic/linear and partially flat dispersions. The significant loop-like charge distribution in the AA stacking can create a pair of linear bands intersecting at $E_F$. The 1D metal becomes a direct semiconductor during the variation of stacking configuration, while the fourfold degeneracy near the zone boundary keeps the same. The metal-semiconductor transition also appears in folded zigzag nanoribbons. On the other side, the AA' and $AB_\beta$ stackings have the nearly flat structures, as observed in monolayer system. There are two relatively stable magnetic configurations: AFM-AFM and AFM-FM. Most of the magnetic bilayer systems have indirect gaps and fourfold degeneracy near $k_x = 1$. As to the interlayer AFM configuration, energy bands are split into the spin-up and spin-down ones under the distinct magnetic environments except for those of the AA' stacking. The 1D and 2D bilayer systems quite differ from each other in terms of the most stable configuration, spatial charge distribution, low-lying energy dispersions, van Hove singularities, band overlap/band gap, magnetism, and spin degeneracy, further illustrating the critical roles of the combined effects.

The edge passivation and the curvature effect could thoroughly change geometric structures and electronic properties of armchair graphene nanoribbons. For the edge-terminated armchair nanoribbons, the former has a very strong effect on adatom arrangements, bond lengths, charge distributions, and energy dispersions. Elements, with an atomic number of less than 20, are classified into three types depending on the optimal geometric structures: (I) the planar structures with the open edges, or the heptagon-pentagon/triangle-pentagon/hexagon edges, (II) the open and edge-buckled profiles and (III) the closed and adatom-stacked-up configurations. Especially, the nitrogen-decorated system is the most stable one with a heptagon-pentagon structure at the edges. The non-monotonous energy dispersions and the adatom-dominated bands come to exist after the adatom decoration, while the weakly dispersive. bands due to the edge carbon atoms disappear. The dependence of energy gap on ribbon width might become more complicated, e.g., the absence of a linearly inverse relationship, and the existence of the semiconductor-metal transition. The atom- and carbon-orbital-projected DOSs are available in identifying the significant C-C, X-C and X-X bondings. On the other side, during the unzipping of an armchair nanotube, three stable geometric structures are revealed in the edge decorations for the elements with an atomic number less than 30. The first type is a regular/deformed zipper-like nanotube formed by the covalent bonds of decorating (B,N,V,Cr,Mn,Fe,Co,Ni,Cu) adatoms. The second one consists of the highly curved zigzag nanoribbons created by the dipole-dipole interactions between two open edges when decorated with Be, Mg, Al and Ti adatoms. The final structure is a flat nanoribbon produced due to the repulsive forces between two edges, and most decorated structures belong to this type. Various decorating adatoms, different curvature angles, and the zigzag edge structure are reflected in the electronic properties, magnetic properties, and bonding configurations. Most of the resulting structures are conductors with relatively high free carrier densities, compared with metallic armchair nanotubes. However, few are semiconductors due to the zigzag-edge-induced anti-ferromagnetism.

The alkali-adsorbed graphene nanoribbons exhibit the feature-rich geometric, electronic and magnetic properties, being dominated by the finite-size confinement, the critical orbital hybridizations in C-C, X-C, and X-X bonds, and the spin arrangements at two edges. The optimal position of adatom is situated at the hollow site of a planar graphene, in which its height grows with the atomic number. The predicted structure could be verified by the STM measurements. The well-extended $\pi$ bonding in graphene is somewhat affected by alkali adsorption; furthermore, there exist charge variations between alkali and carbon/alkali atoms. The former is responsible for the modifications on the carbon-dominated valence bands. The latter arises from the $s$-$2p_z$ and $s$-$s$ orbital hybridizations so that the alkali- and carbon-dependent conduction bands depend on the kind, distribution, and concentration of adatoms. Specially, a linear relation between free electron density and adatom concentration is thoroughly examined from various surface adsorptions. That is, each alkali atom contributes the outmost $s$ orbital to become a conduction electron under any adatom configurations. The creation of very high carrier density indicates that the 1D $n$-type alkali-adsorbed graphene nanoribbons might have potential applications in nano-electronic devices. The metallic alkali-adsorbed zigzag nanoribbons can present three types of spin configurations: the anti-ferromagnetic ordering across two edges, ferromagnetic ordering along one edge and non-magnetism. These are reflected in the edge-localized energy bands and the low-energy prominent peaks of DOS. That the single-edge adatom adsorption leads to the spin-split metallic energy bands could be considered as promising materials for future applications in spintronic devices [380]. There exist important differences between the alkali- and halogen-adsorbed systems in terms of the optimal hollow/top site, the planar/buckled structure, the single-/multi-orbital hybridizations in X-C bonds, the semiconducting/metallic/behavior, and the magnetic/non-magnetic spin configuration. In addition to graphene, the emergent layered materials include silicene, germanene, tinene, phosphorene, $MoS_2$, and so on. Whether the alkali adsorptions on these semiconducting nanoribbon systems could create the high free carrier density is worthy of a thorough and systematic study.

Halogenated GNRs exhibit the diverse and unique chemical bondings, especially for the significant differences between fluorination and other halogenations. The detailed first-principles calculations show that the diverse electronic and magnetic properties mainly come from the critical orbital hybridizations in X-X, X-C, and C-C bonds, and the spin configurations due to the adatoms and zigzag-edge carbons. The important differences between Cl- and F-adsorbed effects cover the adatom-carbon bond length, the planar/buckled honeycomb lattice, the strength of orbital hybridizations, the metallic or semiconducting behavior, the adsorption-dependent hole density, the changes of energy bands, and the adatom-induced spin states. The $(2p_x, 2p_y, 2p_z)$ orbitals of F and C have very strong hybridizations among one another, leading to the buckled structure with the drastic change, especially in $\pi$ bonding. The $(2p_x, 2p_y)$ orbitals of F adatoms can built energy bands, combined with those of C atoms by the F-C bonds. Under the low fluorination, they become narrow (F,C)-co-dominated bands at middle energy. Fluorinated graphenes are hole-doped metals or semiconductors, in which they are very sensitive to the concentration and distribution of adatoms. The F-adsorbed GNRs could create one or two holes per unit cell, depending on F-coverages. On the other hand, only the significant interactions of $p_z$ orbitals exist in other X-C bonds, and the $\sigma$ bonding of GNR is almost unchanged. The X-X bonds are presented at higher concentrations. All of the Cl-related GNRs belong to the p-type metals with FM/NM/AFM configurations, and two/one adatoms provide about one hole per unit cell above/below the 10% concentration. Such critical mechanisms are examined from the atom-dominated band structures, free hole densities, spatial charge densities, net magnetic moment, spin distributions, and orbital- and spin-projected DOSs. The predicted halogenation effects on the optimal geometry, energy bands, and van Hove singularities in DOS could be verified using STM/TEM, ARPES, and STS, respectively. The metallic or semiconducting behaviors with or without magnetism indicate highly potential applications, such as electronic, optical, and spintronic devices. Specifically, the Cl-related GNRs are predicted to exhibit the highest-density free holes with the spin-split FM energy bands.

The metal atadoms, Al/Ti/Bi, can induce the metallic behaviors, being in sharp contrast with the alkali ones. They clearly display the important differences in the essential properties. The Al and Ti guest atoms exhibit the hollow-site optimal positions, while the former might have the $y$-direction shifts, especially for the non-symmetric distributions. The similar shifts are revealed in the deviated bridge-site Bi adatoms. The adatom chemisorptions are relatively easily observed in the Ti chemisorptions with the largest binding energies; that is, the Ti-adatom adsorption has the highest concentration. Energy bands, which crossing the Fermi level, mainly arise from carbon atoms (metal atoms) for Al-adsorbed systems (Ti- and Al-doped ones). There exist the Al-dominated valence bands at $E^v \sim 3$ eV and the partial adatom contributions to the conduction bands, being consistent with the $3p_z - 2p_z$ and $(3p_x + 3p_y) - 2p_z$ orbital hybridizations in Al-C bonds. This feature is also supported by the spatial charge distributions/DOSs. At higher adatom concentrations, the significant $3p_x - 3p_x$ & $3p_y - 3p_y$ hybridizations in Al-Al bonds are indicated by the enhanced Al-dependent conduction bands/DOSs with the wider energy widths. The multi-orbital hybridizations in Ti-C bonds are very complicated, in which five orbitals of adatom, $(d_{xy}, d_{xz}, d_{yz}, d_{z^2}$, and $d_{x^2-y^2}$ have the strong interactions with $2p_z$ orbitals of carbons. Such orbitals also take part in the Ti-Ti chemical bondings. It is very difficult to observe more complex orbital interactions in the other condensed-matter systems. Three kinds of observable Bi-C bondings, $6p_z - 2p_z$, $6p_x - 2p_z$ & $6p_y - 2p_z$, are characterized by the charge distributions/DOSs except for the symmetric single-adatom adsorptions. The important $(6p_x, 6p_y, 6p_z) - (6p_x, 6p_y, 6p_z)$

orbital hybridizations in Bi-Bi bonds are responsible for the Bi-induced low-lying energy bands. As to the magnetic properties, the Al adatoms do not create the spin distributions, but their interactions with the zigzag carbon atoms can destroy the latter's ones and thus create the FM or NM configurations. Armchair and zigzag Al-adsorbed graphene naoribbons, respectively, belong to the NM and AFM/FM/NM metals. On the other side, most of Ti- and Bi-adsorbed asymmetric systems exhibit the FM configurations under the themself-induced spin states and the greatly reduced edge-carbon magnetic moments. Furthermore, the symmetric adatom distributions in zigzag systems might lead to the very complicated AFM configurations, being in sharp contrast with the pristine magnetic configuration. The AFM spin arrangements is strongly associated with the edge carbons and adsorbates. For the Ti adatoms, the strength of magnetic response only relies on the concentration for the double-side adsorptions. The Fe/Co/Ni-adsorbed systems exhibit similar optimal geometric configurations compare to the Ti ones, and the Fe/Co adatoms also create the spin split electronic structures in both armchair and zigzag systems. However, Ni adatom being a special transition metal which do not create the spin distributions.

This review work could serve as a first step toward a full understanding of other layered condensed-matter systems characterized by the nano-scaled thickness and width, the specific lattice symmetries, and the different stacking configurations. In addition to graphene/graphene nanoribbon, the emergent layered materials have been successfully synthesized by the various experimental methods. For example, the 1D silicene, germanene, stanene, phosphorene, and $MoS_2$ nanoribbons are, respectively, grown Ag(111)/Ir(111)/ZrB2 surfaces [85, 82, 399, 199, 196]. They are expected to be suitable for exploring the novel physical, chemical and material properties and have high potentials in near-future applications. Such main-stream materials possess the unique geometric structures and the rich intrinsic interactions, covering the atom-dependent lattice symmetries, the planar/buckled/puckered/curved structures, the achiral/chiral edge structures, the diversified stacking configurations, the equivalent/inequivalent sublattices, the intra- and inter-layer atomic interactions, the single- and multi-orbital chemical bondings, the atom- and orbital-induced site energies, two kinds of spin-orbital couplings, and the atom- and edge-dependent spin configurations. The composite effects will be directly revealed in the various Hamiltonians. Up to now, there are some theoretical calculations on the essential properties using the first-principles method [192, 215] and the tight-binding model [198, 23], illustrating the unusual geometric structures, electronic properties and magnetic configurations. According to the current theoretical framework, the former could provide very thorough and systematic investigations on the edge-, buckling-, folding-, curving-, scrolling-, layer-, stacking-, edge-decoration-, and adatom-absorption-created diverse phenomena. The diversities among the similar/distinct layered materials are one of the studying focuses, being distinguished from the critical characteristics of the orbital bondings and spin configurations. As to the quantization phenomena arising from the external magnetic field, the latter is available in exploring the main features of Landau levels, and the magneto-optical selection rules.

Part of theoretical calculations are consistent with the experimental measurements. The STM/TEM/AFM/SEM/XPS/x-ray diffraction measurements are utilized to confirm the feature-rich geometric structures in various pristine/edge-terminated graphene nanoribbons, covering the armchair/zigzag/chiral open boundaries [290], the curved surfaces (the unzipped nanotubes) [166], the bilayer-like folded structures (the composite nanoribbon and nanotube structures) [193], the spiral nanoscrolls [345], the AB/AA stackings [266], and the H/Cl/O decorations with the planar or distorted edge structures [390]. They are available in examining the optimal geometries of adatom-adsorbed systems, such as, the adsorption positions, the X-C and X-X bond lengths, and the planar/buckled honeycomb lattices. As to electronic structures of planar graphene nanoribbons, the ARPES measurements have confirmed the 1D parabolic energy dispersions near the $\Gamma$ point, accompanied with as an energy gap and distinct energy spacings [294]. The structure- and adatom-induced drastic changes in energy bands, especially for the metallic/semiconducting behaviors, the dominance of carbons and adatoms, and the spin splitting/degeneracy, are worthy of detailed experimental investigations. The width- and edge-dependent energy gaps and the 1D asymmetric peaks in DOSs have been verified for graphene nanoribbons using the STS measurements [336]. The further STS/SP-STS examinations are required to observe the diverse electronic and magnetic properties, the FM/AFM/NM semiconductors and metals due to geometric symmetries, passivation/decoration, and adsorption. The rich and unique essential properties will be further revealed in transport and optical properties. The transport experiments could test the semiconducting or metallic behavior, the n- or p-type doing, the relation between the conduction electron/valence hole density and the adatom concentration, and the spin-polarized currents, such as the Hall and Seebeck effects [372]. Moreover, the optical/magneto-optical experiments are useful in identifying the main features of the spin-split band structure, e.g., the Kerr effects [94, 67].

The diverse essential properties clearly indicate that the structure- and adatom-enriched graphene nanoribbons are highly potential materials in the near-future applications. The pristine graphene nanoribbons, with planar/layered structures and semiconducting characteristics, could serve as FETs [355], chemical/biochemical sensors [223], and lithium-ion battery anode [204], The folded graphene nanoribbons are suitable for FETs [193], and high-performance flexible electrode materials [219]. Carbon nanoscrolls possess the hollow spiral structure, the unusual electronic properties, and the special mechanical strains, so that they could be utilized for hydrogen storages [246], microcircuit interconnect components [370], and electromechanical nanoactuators [295]. Specifically, the tunable metallic and semiconducting behaviors of fluorinated graphene nanoribbons, as identified from 2D systems, are expected to have various applications, e.g., lithium batteries [144], supercapacitors [11], molecule detectors [306], and biosensors [343]. Moreover, the alkali-, Cl- and metal-adsorbed graphene nanoribbons, which could generate very high-density free carriers in the presence/absence of magnetism, are promising candidates in developing the next-generation electronic/spintronic devices [46], high-capacity batteries [205], and energy storages [279].

The essential properties of graphene nanoribbons are easily modulated by the electric and magnetic fields, and the uniaxial stress, such as electronic properties and optical absorption spectra. Based on the tight-binding model calculations, the distinct Coulomb potential energies on lattice sites and intralayer/interlayer hopping integrals of the neighboring atoms might drastically alter the band-edge states, energy gap, subband spacings, dependence of wave vector, state degeneracy and crossing/non-crossing/anticrossing behaviors [202]. For monolayer systems, the transverse electric field and the mechanical deformation would induce the semiconductor-metal transitions [202], being worthy of experimental examinations by using the STS, transport and optical measurements. A uniform perpendicular electric field in few-layer graphene nanoribbons is also an effective method in creating the dramatic transition, in which the intrinsic systems might be metals, if the interlayer hopping integrals could suppress the quantum confinement, Specially, the neighboring electronic states, with close energies, are magnetically quantized into the 1D Landau subbands composed of the partially flat and parabolic energy dispersions [75]. The well-behaved standing waves become the localized Landau wave functions with the specific oscillatory modes. The highly degenerate states are initiated from the Fermi level. The energy range, which they could survive, depends on the strong competition between the magnetic quantization and kinetic energy. There are a lot of delta-function-like symmetric peaks in the low-lying DOSs. However, the magnetic field cannot lead to obvious changes in energy gaps. It should be noticed that a perpendicular composite electric and magnetic field is expected to destroy state degeneracy and generate the frequently anti-crossing/crossing phenomena, as observed in few-layer graphenes [202]. Concerning optical properties, the H-passivated graphene nanoribbons exhibit the edge-dependent selection rules, $\Delta J = 0$ and $|\Delta J| = 2I + 1$ for armchair and zigzag systems, respectively. Such rules arise from the characteristics of the normal standing waves in valence and conduction states. On the other hand, the magneto-optical selection rule is $|\Delta n| = 1$, regardless of the edge structures. In general, a plenty of prominent absorption peaks are, respectively, revealed in the symmetric and square-root asymmetric forms at low and high energies. This clearly illustrates the competitive relation arising from the significant edge structure and $B_z$-field.

## Problems

The following problems are solved by the generalized tight-binding model, the Hubbard model and the first-principles method.

**(1)** For H-passivated monolayer graphene nanoribbons, the low-energy electronic properties are qualitatively approximated by ignoring the very strong C-H bonds (the deeper energy states). (a) Calculate the analytic Hamiltonian of the nearest-neighbor $2p_z$ orbitals by using the tight-binding model for armchair and zigzag systems. (b) Plot valence and conduction bands of $N_A = 8, 9$, and 10 and $N_z = 4, 5$, and 6. (c) Evaluate densities of states (DOSs) for the above-mentioned systems. (d) Discuss the width-dependent energy gaps, especially for three types of armchair graphene nanoribbons.

**(2)** When the H-terminated monolayer systems exist in a transverse uniform electric field $(E_y \hat{y})$, the effective Hamiltonian has extra site energies due to the position-dependent Coulomb potential energies. (a) Investigate the drastic changes of electronic properties, including the energy spacings of the band-edge states, band gaps and energy disperions in armchair and zigzag graphene nanoribbons. (b) Explore the semiconductor-metal transitions during the variation of $E_y$.

**(3)** Similar to problem in (2), but the 1D systems under an uniaxial stress. The obvious changes in the hopping intergrals are assumed to be inversely proportional to the square of interaction lengths (the Harrison's rule). The non-uniform atomic interactions will dominate the essential properties, such as the oscillatory relations between the semiconductor-metal transition and the mechanical strain.

**(4)** A uniform perpendicular magnetic field $(B_z \hat{z})$ creates the vector potential and thus the periodical Peierls phase. (a) Evaluate the magnetic Hamiltonian formulism in problem (1). (b) By the numerical calculations, plot the Landau subbands composed of the partially flat and parabolic dispersions. (c) Evaluate the $B_z$ dependence for the magnetic energy spectra. (d) Discuss valence and conduction wave functions in the presence and absence of $B_z$, and then define a magnetic quantum number from the Landau wave functions.

**(5)** Discuss the important differences between 1D armchair/zigzag graphene nanoribbons and 2D monolayer graphene in magneto-electronic properties, covering the $B_z$-dependent energy spectra and wave functions, i.e., explore the relations between the magnetic quantization and quantum confinement.

**(6)** Consider AA and AB bilayer graphene nanoribbons with hydrogen terminations. (a) Similar to problem (1), and (b), and investigate the effects on the electronic properties arising from a transverse/perpendicular electric field, and a perpendicular magnetic field.

**(7)** When the H-passivated monolayer graphene nanoribbons are present in an electromagnetic wave, the occupied valence states are excited to the unoccupied ones under the conservation of energy and momentum and the requirement of the Fermi-Dirac distributions. The Kubo formulism under the gradient approximation could deal with the electric-dipole optical excitations. (a) Calculate the absorption spectra and investigate the edge-dependent selection rules. (b) Discuss the drastic changes due to the external electric and magnetic fields and the uniaxial deformations.

**(8)** The 1D systems might exhibit the plateau structures in the ballistic transport. By using the Landauer-Buttiker formulism, the dependence of electrical conductances on (a) edge structures, (b) nanoribbon widths, (c) transverse electric fields, and (d) perpendicular magnetic fields could be studied thoroughly.

**(9)** Hydrogen atoms are easily absorbed on the surfaces of graphene nanoribbons in the experimental syntheses. (a) Determine the optimal geometric structure based on the first-principles calculations of the VASP, including the non-planar honeycomb lattice and the C-H bond length accompanied with the distinct C-C bond lengths. The critical orbital hybridizations in the significant C-H bonds could be identified from (b) the atom-dominated energy bands, (c) the spatial charge distributions and (d) the atom- and orbital-decomposed DOSs. Furthermore, the ferromagnetic and anti-ferromagnetic spin configurations might be induced by H adatoms and zigzag edge carbons. Plot (e) the spin-split energy bands and the spin density distributions, and explore the dependences on adatom positions and edge structures. Specifically, investigate (f) the relations (conditions) between adatom concentrations and spin polarizations.

**(10)** Oxidization of graphene nanoribbons will induce the dramatic change in electronic properties. Discuss the similar problems (a)–(d) of (9). It should be noticed that the magnetic properties are expected to be absent according to the theoretical predictions on oxygen-adsorbed graphenes.

**(11)** Try to find a metal-absorbed graphene nanoribbon with the maximum $n$-type free carrier density from the periodic table. The 1D linear electron density is exactly determined from the Fermi momenta of energy bands. It would be helpful in understanding the metal-atom doping behaviors of the graphene-related systems.

**(12)** Add $FeCl_3$, $HNO_3$ and $H_2SO_4$ in graphene nanoribbons. Explore whether the functional molecules will induce the $p$-type dopings, the dramatic changes in the main features of electronic properties, and the spin configurations.

# References

[1] A. N. Abbas, G. Liu, B. Liu, L. Zhang, H. Liu, D. Ohlberg, W. Wu, and C. Zhou. Patterning, characterization, and chemical sensing applications of graphene nanoribbon arrays down to 5 nm using helium ion beam lithography. *ACS Nano*, 8(2):1538–1546, 2014.

[2] A. Ahmadi Peyghan, N. L. Hadipour, and Z. Bagheri. Effects of al doping and double-antisite defect on the adsorption of hcn on a bc2n nanotube: density functional theory studies. *The Journal of Physical Chemistry C*, 117(5):2427–2432, 2013.

[3] H. Ajiki and T. Ando. Magnetic properties of carbon nanotubes. *Journal of the Physical Society of Japan*, 62(7):2470–2480, 1993.

[4] H. Ajiki and T. Ando. Aharonov-bohm effect in carbon nanotubes. *Physica B: Condensed Matter*, 201:349–352, 1994.

[5] H. Ajiki and T. Ando. Magnetic properties of ensembles of carbon nanotubes. *Journal of the Physical Society of Japan*, 64(11):4382–4391, 1995.

[6] H. Ajiki and T. Ando. Energy bands of carbon nanotubes in magnetic fields. *Journal of the Physical Society of Japan*, 65(2):505–514, 1996.

[7] N. Akima, Y. Iwasa, S. Brown, A. M. Barbour, J. Cao, J. L. Musfeldt, H. Matsui, N. Toyota, M. Shiraishi, and H. Shimoda. Strong anisotropy in the far-infrared absorption spectra of stretch-aligned single-walled carbon nanotubes. *Advanced Materials*, 18(9):1166–1169, 2006.

[8] O. Ü. Aktürk and M. Tomak. Bismuth doping of graphene. *Applied Physics Letters*, 96(8):081914, 2010.

[9] J. Alfonsi and M. Meneghetti. Excitonic properties of armchair graphene nanoribbons from exact diagonalization of the hubbard model. *New J. Phys.*, 14(5):053047, 2012.

[10] H. Amara, S. Latil, V. Meunier, P. Lambin, and J.-C. Charlier. Scanning tunneling microscopy fingerprints of point defects in graphene: A theoretical prediction. *Physical Review B*, 76(11):115423, 2007.

[11] H. An, Y. Li, P. Long, Y. Gao, C. Qin, C. Cao, Y. Feng, and W. Feng. Hydrothermal preparation of fluorinated graphene hydrogel for high-performance supercapacitors. *Journal of Power Sources*, 312:146–155, 2016.

[12] Z. Ao, Q. Jiang, R. Zhang, T. Tan, and S. Li. Al doped graphene: a promising material for hydrogen storage at room temperature. *Journal of Applied Physics*, 105(7):074307, 2009.

[13] Z. Ao, J. Yang, S. Li, and Q. Jiang. Enhancement of co detection in al doped graphene. *Chemical Physics Letters*, 461(4-6):276–279, 2008.

[14] G. Autès and O. V. Yazyev. Engineering quantum spin hall effect in graphene nanoribbons via edge functionalization. *Physical Review B*, 87(24):241404, 2013.

[15] A. Bachtold, C. Strunk, J.-P. Salvetat, J.-M. Bonard, L. Forró, T. Nussbaumer, and C. Schönenberger. Aharonov–bohm oscillations in carbon nanotubes. *Nature*, 397(6721):673–675, 1999.

[16] R. Bacon. Growth, structure, and properties of graphite whiskers. *Journal of Applied Physics*, 31(2):283–290, 1960.

[17] J. Bai, X. Duan, and Y. Huang. Rational fabrication of graphene nanoribbons using a nanowire etch mask. *Nano Letters*, 9(5):2083–2087, 2009.

[18] K.-K. Bai, Y. Zhou, H. Zheng, L. Meng, H. Peng, Z. Liu, J.-C. Nie, and L. He. Creating one-dimensional nanoscale periodic ripples in a continuous mosaic graphene monolayer. *Physical Review Letters*, 113(8):086102, 2014.

[19] R. Balog, B. Jørgensen, L. Nilsson, M. Andersen, E. Rienks, M. Bianchi, M. Fanetti, E. Lægsgaard, A. Baraldi, and S. Lizzit. Bandgap opening in graphene induced by patterned hydrogen adsorption. *Nature Materials*, 9(4):315–319, 2010.

[20] S. Bandow, S. Asaka, Y. Saito, A. Rao, L. Grigorian, E. Richter, and P. Eklund. Effect of the growth temperature on the diameter distribution and chirality of single-wall carbon nanotubes. *Physical Review Letters*, 80(17):3779, 1998.

[21] V. Barone, O. Hod, and G. E. Scuseria. Electronic structure and stability of semiconducting graphene nanoribbons. *Nano Letters*, 6(12):2748–2754, 2006.

[22] L. X. Benedict, V. H. Crespi, S. G. Louie, and M. L. Cohen. Static conductivity and superconductivity of carbon nanotubes: Relations between tubes and sheets. *Physical Review B*, 52(20):14935, 1995.

[23] G. Berdiyorov, M. Neek-Amal, F. Peeters, and A. C. van Duin. Stabilized silicene within bilayer graphene: A proposal based on molecular dynamics and density-functional tight-binding calculations. *Physical Review B*, 89(2):024107, 2014.

[24] O. L. Berman, G. Gumbs, and Y. E. Lozovik. Magnetoplasmons in layered graphene structures. *Physical Review B*, 78(8):085401, 2008.

[25] D. Bernaerts, A. Zettl, N. G. Chopra, A. Thess, and R. Smalley. Electron diffraction study of single-wall carbon nanotubes. *Solid State Communications*, 105(3):145–149, 1998.

[26] J. Bernal. The structure of graphite. *Proceedings of the Royal Society of London. Series A, Containing Papers of a Mathematical and Physical Character*, 106(740):749–773, 1924.

[27] T. Bhardwaj, A. Antic, B. Pavan, V. Barone, and B. D. Fahlman. Enhanced electrochemical lithium storage by graphene nanoribbons. *Journal of the American Chemical Society*, 132(36):12556–12558, 2010.

[28] S. Bhattacharyya and A. K. Singh. Lifshitz transition and modulation of electronic and transport properties of bilayer graphene by sliding and applied normal compressive strain. *Carbon*, 99:432–438, 2016.

[29] G. Binnig and H. Rohrer. Scanning tunnel microscope. *Helvetica Physica Acta*, 55(2):726–735, 1982.

[30] G. Binnig and H. Rohrer. Scanning tunneling microscopy from birth to adolescence. *Reviews of Modern Physics*, 59(3):615, 1987.

[31] M. Birowska, K. Milowska, and J. Majewski. Van der waals density functionals for graphene layers and graphite. *Acta Phys Pol A*, 120:845–848, 2011.

[32] M. Black, Y.-M. Lin, S. Cronin, O. Rabin, and M. Dresselhaus. Infrared absorption in bismuth nanowires resulting from quantum confinement. *Physical Review B*, 65(19):195417, 2002.

[33] S. Blankenburg, J. Cai, P. Ruffieux, R. Jaafar, D. Passerone, X. Feng, K. Müllen, R. Fasel, and C. A. Pignedoli. Intraribbon heterojunction formation in ultranarrow graphene nanoribbons. *Acs Nano*, 6(3):2020–2025, 2012.

[34] X. Blase, L. X. Benedict, E. L. Shirley, and S. G. Louie. Hybridization effects and metallicity in small radius carbon nanotubes. *Physical Review Letters*, 72(12):1878, 1994.

[35] P. E. Blöchl. Projector augmented-wave method. *Physical Review B*, 50(24):17953, 1994.

[36] S. Bobaru, É. Gaudry, M.-C. De Weerd, J. Ledieu, and V. Fournée. Competing allotropes of bi deposited on the al 13 co 4 (100) alloy surface. *Physical Review B*, 86(21):214201, 2012.

[37] A. Bostwick. Coexisting massive and massless dirac fermions in symmetry-broken bilayer graphene. *Bulletin of the American Physical Society*, 59, 2014.

[38] A. Bostwick, T. Ohta, T. Seyller, K. Horn, and E. Rotenberg. Quasiparticle dynamics in graphene. *Nature Physics*, 3(1):36–40, 2007.

[39] C. Bower, S. Suzuki, K. Tanigaki, and O. Zhou. Synthesis and structure of pristine and alkali-metal-intercalated single-walled carbon nanotubes. *Applied Physics A: Materials Science & Processing*, 67(1):47–52, 1998.

[40] S. F. Braga, V. R. Coluci, S. B. Legoas, R. Giro, D. S. Galvão, and R. H. Baughman. Structure and dynamics of carbon nanoscrolls. *Nano Letters*, 4(5):881–884, 2004.

[41] L. Brey and H. Fertig. Elementary electronic excitations in graphene nanoribbons. *Physical Review B*, 75(12):125434, 2007.

[42] M. Burghard, H. Klauk, and K. Kern. Carbon-based field-effect transistors for nanoelectronics. *Advanced Materials*, 21(25-26):2586–2600, 2009.

[43] L. Bursill, P. A. Stadelmann, J. Peng, and S. Prawer. Surface plasmon observed for carbon nanotubes. *Physical Review B*, 49(4):2882, 1994.

[44] J. Cai, P. Ruffieux, R. Jaafar, M. Bieri, T. Braun, S. Blankenburg, M. Muoth, A. P. Seitsonen, M. Saleh, and X. Feng. Atomically precise bottom-up fabrication of graphene nanoribbons. *Nature*, 466(7305):470–473, 2010.

[45] J. Campos-Delgado, J. M. Romo-Herrera, X. Jia, D. A. Cullen, H. Muramatsu, Y. A. Kim, T. Hayashi, Z. Ren, D. J. Smith, and Y. Okuno. Bulk production of a new form of sp$^2$ carbon: crystalline graphene nanoribbons. *Nano Letters*, 8(9):2773–2778, 2008.

[46] A. Candini, S. Klyatskaya, M. Ruben, W. Wernsdorfer, and M. Affronte. Graphene spintronic devices with molecular nanomagnets. *Nano Letters*, 11(7):2634–2639, 2011.

[47] A. G. Cano-Márquez, F. J. Rodríguez-Macías, J. Campos-Delgado, C. G. Espinosa-González, F. Tristán-López, D. Ramírez-González, D. A. Cullen, D. J. Smith, M. Terrones, and Y. I. Vega-Cantú. Ex-mwnts: graphene sheets and ribbons produced by lithium intercalation and exfoliation of carbon nanotubes. *Nano Letters*, 9(4):1527–1533, 2009.

[48] J. Cao, Q. Wang, and H. Dai. Electron transport in very clean, as-grown suspended carbon nanotubes. *Nature Materials*, 4(10):745–749, 2005.

[49] J. Cao, Q. Wang, M. Rolandi, and H. Dai. Aharonov-bohm interference and beating in single-walled carbon-nanotube interferometers. *Physical Review Letters*, 93(21):216803, 2004.

[50] M. Caragiu and S. Finberg. Alkali metal adsorption on graphite: a review. *Journal of Physics: Condensed Matter*, 17(35):R995, 2005.

[51] F. Cataldo, G. Compagnini, G. Patané, O. Ursini, G. Angelini, P. R. Ribic, G. Margaritondo, A. Cricenti, G. Palleschi, and F. Valentini. Graphene nanoribbons produced by the oxidative unzipping of single-wall carbon nanotubes. *Carbon*, 48(9):2596–2602, 2010.

[52] F. Cervantes-Sodi, G. Csanyi, S. Piscanec, and A. Ferrari. Edge-functionalized and substitutionally doped graphene nanoribbons: Electronic and spin properties. *Physical Review B*, 77(16):165427, 2008.

[53] J. Červenka, M. Katsnelson, and C. Flipse. Room-temperature ferromagnetism in graphite driven by two-dimensional networks of point defects. *Nature Physics*, 5(11):840–844, 2009.

[54] K. T. Chan, J. Neaton, and M. L. Cohen. First-principles study of metal adatom adsorption on graphene. *Physical Review B*, 77(23):235430, 2008.

[55] C. Chang, B. Wu, R. Chen, and M.-F. Lin. Deformation effect on electronic and optical properties of nanographite ribbons. *Journal of Applied Physics*, 101(6):063506, 2007.

[56] C.-P. Chang, Y.-C. Huang, C. Lu, J.-H. Ho, T.-S. Li, and M.-F. Lin. Electronic and optical properties of a nanographite ribbon in an electric field. *Carbon*, 44(3):508–515, 2006.

[57] S.-L. Chang, S.-Y. Lin, S.-K. Lin, C.-H. Lee, and M.-F. Lin. Geometric and electronic properties of edge-decorated graphene nanoribbons. *Scientific Reports*, 4:6038, 2014.

[58] S.-L. Chang, B.-R. Wu, J.-H. Wong, and M.-F. Lin. Configuration-dependent geometric and electronic properties of bilayer graphene nanoribbons. *Carbon*, 77:1031–1039, 2014.

[59] S.-L. Chang, B.-R. Wu, P.-H. Yang, and M.-F. Lin. Curvature effects on electronic properties of armchair graphene nanoribbons without passivation. *Physical Chemistry Chemical Physics*, 14(47):16409–16414, 2012.

[60] S.-L. Chang, B.-R. Wu, P.-H. Yang, and M. F. Lin. Geometric, magnetic and electronic properties of folded graphene nanoribbons. *RSC Advances*, 6(69):64852–64860, 2016.

[61] P. Chantharasupawong, R. Philip, N. T. Narayanan, P. M. Sudeep, A. Mathkar, P. M. Ajayan, and J. Thomas. Optical power limiting in fluorinated graphene oxide: an insight into the nonlinear optical properties. *The Journal of Physical Chemistry C*, 116(49):25955–25961, 2012.

[62] J.-C. Charlier, H. Amara, and P. Lambin. Catalytically assisted tip growth mechanism for single-wall carbon nanotubes. *ACS Nano*, 1(3):202–207, 2007.

[63] H.-H. Chen, S. Su, S.-L. Chang, B.-Y. Cheng, S. Chen, H.-Y. Chen, M. F. Lin, and J. Huang. Tailoring low-dimensional structures of bismuth on monolayer epitaxial graphene. *Scientific Reports*, 5, 2015.

[64] H.-H. Chen, S. Su, S.-L. Chang, B.-Y. Cheng, C.-W. Chong, J. Huang, and M. F. Lin. Long-range interactions of bismuth growth on monolayer epitaxial graphene at room temperature. *Carbon*, 93:180–186, 2015.

[65] J. H. Chen, C. Jang, S. Adam, M. Fuhrer, E. Williams, and M. Ishigami. Charged impurity scattering in graphene. *Nature Physics*, 4:377381, 2008.

[66] J.-W. Chen, H.-C. Huang, D. Convertino, C. Coletti, L.-Y. Chang, H.-W. Shiu, C.-M. Cheng, M.-F. Lin, S. Heun, and F. S.-S. Chien. Efficient n-type doping in epitaxial graphene through strong lateral orbital hybridization of ti adsorbate. *Carbon*, 109:300–305, 2016.

[67] J.-Y. Chen, J. Zhu, D. Zhang, D. M. Lattery, M. Li, J.-P. Wang, and X. Wang. Time-resolved magneto-optical kerr effect of magnetic thin films for ultrafast thermal characterization. *The Journal of Physical Chemistry Letters*, 7(13):2328–2332, 2016.

[68] S.-C. Chen, J.-Y. Wu, and M.-F. Lin. Novel magnetic quantization of bismuthene. *arXiv preprint arXiv:1709.03289*, 2017.

[69] Y. Chen, J. Lu, and Z. Gao. Structural and electronic study of nanoscrolls rolled up by a single graphene sheet. *The Journal of Physical Chemistry C*, 111(4):1625–1630, 2007.

[70] Y.-C. Chen, D. G. De Oteyza, Z. Pedramrazi, C. Chen, F. R. Fischer, and M. F. Crommie. Tuning the band gap of graphene nanoribbons synthesized from molecular precursors. *ACS Nano*, 7(7):6123–6128, 2013.

[71] Z. Chen and X.-Q. Wang. Stacking-dependent optical spectra and many-electron effects in bilayer graphene. *Physical Review B*, 83(8):081405, 2011.

[72] V. Cherkez, G. T. de Laissardière, P. Mallet, and J.-Y. Veuillen. Van hove singularities in doped twisted graphene bilayers studied by scanning tunneling spectroscopy. *Physical Review B*, 91(15):155428, 2015.

[73] M. Chi and Y.-P. Zhao. Adsorption of formaldehyde molecule on the intrinsic and al-doped graphene: a first principle study. *Computational Materials Science*, 46(4):1085–1090, 2009.

[74] S. W. Chu, S. J. Baek, D. C. Kim, S. Seo, J. S. Kim, and Y. W. Park. Charge transport in graphene doped with diatomic halogen molecules (i 2, br 2) near dirac point. *Synthetic Met.*, 162(17):1689–1693, 2012.

[75] H.-C. Chung, C.-P. Chang, C.-Y. Lin, and M. F. Lin. Electronic and optical properties of graphene nanoribbons in external fields. *Physical Chemistry Chemical Physics*, 18(11):7573–7616, 2016.

[76] L. Ci, Z. Xu, L. Wang, W. Gao, F. Ding, K. F. Kelly, B. I. Yakobson, and P. M. Ajayan. Controlled nanocutting of graphene. *Nano Research*, 1(2):116–122, 2008.

[77] C. Coletti, S. Forti, A. Principi, K. V. Emtsev, A. A. Zakharov, K. M. Daniels, B. K. Daas, M. Chandrashekhar, T. Ouisse, and D. Chaussende. Revealing the electronic band structure of trilayer graphene on sic: An angle-resolved photoemission study. *Physical Review B*, 88(15):155439, 2013.

[78] J. Cowley, P. Nikolaev, A. Thess, and R. E. Smalley. Electron nano-diffraction study of carbon single-walled nanotube ropes. *Chemical Physics Letters*, 265(3-5):379–384, 1997.

[79] T. Cuk, D. Lu, X. Zhou, Z.-X. Shen, T. Devereaux, and N. Nagaosa. A review of electron–phonon coupling seen in the high-tc superconductors by angle-resolved photoemission studies (arpes). *Physica Status Solidi (b)*, 242(1):11–29, 2005.

[80] C. G. da Rocha, P. A. Clayborne, P. Koskinen, and H. Häkkinen. Optical and electronic properties of graphene nanoribbons upon adsorption of ligand-protected aluminum clusters. *Physical Chemistry Chemical Physics*, 16(8):3558–3565, 2014.

[81] S. S. Datta, D. R. Strachan, S. M. Khamis, and A. C. Johnson. Crystallographic etching of few-layer graphene. *Nano Letters*, 8(7):1912–1915, 2008.

[82] M. Dávila, L. Xian, S. Cahangirov, A. Rubio, and G. Le Lay. Germanene: a novel two-dimensional germanium allotrope akin to graphene and silicene. *New Journal of Physics*, 16(9):095002, 2014.

[83] W. A. de Heer, W. Bacsa, A. Chatelain, T. Gerfin, and R. Humphrey-Baker. Aligned carbon nanotube films: production and optical and electronic properties. *Science*, 268(5212):845, 1995.

[84] S. De Jong, E. van Heumen, S. Thirupathaiah, R. Huisman, F. Massee, J. Goedkoop, R. Ovsyannikov, J. Fink, H. Dürr, and A. Gloskovskii. Droplet-like fermi surfaces in the anti-ferromagnetic phase of eufe2as2, an fe-pnictide superconductor parent compound. *EPL (Europhysics Letters)*, 89(2):27007, 2010.

[85] P. De Padova, P. Perfetti, B. Olivieri, C. Quaresima, C. Ottaviani, and G. Le Lay. 1d graphene-like silicon systems: silicene nano-ribbons. *Journal of Physics: Condensed Matter*, 24(22):223001, 2012.

[86] P. De Padova, C. Quaresima, C. Ottaviani, P. M. Sheverdyaeva, P. Moras, C. Carbone, D. Topwal, B. Olivieri, A. Kara, and H. Oughaddou. Evidence of graphene-like electronic signature in silicene nanoribbons. *Applied Physics Letters*, 96(26):261905, 2010.

[87] A. V. De Parga, F. Calleja, B. Borca, M. Passeggi Jr, J. Hinarejos, F. Guinea, and R. Miranda. Periodically rippled graphene: growth and spatially resolved electronic structure. *Physical Review Letters*, 100(5):056807, 2008.

[88] F. Donati, Q. Dubout, G. Autès, F. Patthey, F. Calleja, P. Gambardella, O. Yazyev, and H. Brune. Magnetic moment and anisotropy of individual co atoms on graphene. *Physical Review Letters*, 111(23):236801, 2013.

[89] M. Dresselhaus and G. Dresselhaus. Intercalation compounds of graphite. *Advances in Physics*, 30(2):139–326, 1981.

[90] T. Dürkop, S. Getty, E. Cobas, and M. Fuhrer. Extraordinary mobility in semiconducting carbon nanotubes. *Nano Letters*, 4(1):35–39, 2004.

[91] S. Dutta, A. K. Manna, and S. K. Pati. Intrinsic half-metallicity in modified graphene nanoribbons. *Physical Review Letters*, 102(9):096601, 2009.

[92] G. A. Elia, K. Marquardt, K. Hoeppner, S. Fantini, R. Lin, E. Knipping, W. Peters, J.-F. Drillet, S. Passerini, and R. Hahn. An overview and future perspectives of aluminum batteries. *Advanced Materials*, 28(35):7564–7579, 2016.

[93] A. L. Elías, A. R. Botello-Méndez, D. Meneses-Rodríguez, V. Jehová González, D. Ramírez-González, L. Ci, E. Muñoz-Sandoval, P. M. Ajayan, H. Terrones, and M. Terrones. Longitudinal cutting of pure and doped carbon nanotubes to form graphitic nanoribbons using metal clusters as nanoscalpels. *Nano Letters*, 10(2):366–372, 2009.

[94] C. T. Ellis, A. V. Stier, M.-H. Kim, J. G. Tischler, E. R. Glaser, R. L. Myers-Ward, J. L. Tedesco, C. R. Eddy, D. K. Gaskill, and J. Cerne. Magneto-optical fingerprints of distinct graphene multilayers using the giant infrared kerr effect. *Scientific Reports*, 3, 2013.

[95] S. Fan, M. G. Chapline, N. R. Franklin, T. W. Tombler, A. M. Cassell, and H. Dai. Self-oriented regular arrays of carbon nanotubes and their field emission properties. *Science*, 283(5401):512–514, 1999.

[96] R. Farghadan and E. Saievar-Iranizad. Spin-polarized transport in zigzag-edge graphene nanoribbon junctions. *Journal of Applied Physics*, 111(1):014304, 2012.

[97] J. Feng, L. Qi, J. Y. Huang, and J. Li. Geometric and electronic structure of graphene bilayer edges. *Physical Review B*, 80(16):165407, 2009.

[98] X. Feng, S. Kwon, J. Y. Park, and M. Salmeron. Superlubric sliding of graphene nanoflakes on graphene. *ACS Nano*, 7(2):1718–1724, 2013.

[99] A. L. Friedman, O. M. vant Erve, J. T. Robinson, K. E. Whitener Jr, and B. T. Jonker. Hydrogenated graphene as a homoepitaxial tunnel barrier for spin and charge transport in graphene. *ACS Nano*, 9(7):6747–6755, 2015.

[100] S. Fujii and T. Enoki. Cutting of oxidized graphene into nanosized pieces. *Journal of the American Chemical Society*, 132(29):10034–10041, 2010.

[101] M. Fujita, M. Igami, and K. Nakada. Lattice distortion in nanographite ribbons. *Journal of the Physical Society of Japan*, 66(7):1864–1867, 1997.

[102] M. Fujita, K. Wakabayashi, K. Nakada, and K. Kusakabe. Peculiar localized state at zigzag graphite edge. *Journal of the Physical Society of Japan*, 65(7):1920–1923, 1996.

[103] A. K. Geim and K. S. Novoselov. The rise of graphene. *Nature Materials*, 6(3):183–191, 2007.

[104] G. Giovannetti, P. Khomyakov, G. Brocks, V. v. Karpan, J. Van den Brink, and P. J. Kelly. Doping graphene with metal contacts. *Physical review letters*, 101(2):026803, 2008.

[105] N. Gorjizadeh, A. A. Farajian, K. Esfarjani, and Y. Kawazoe. Spin and band-gap engineering in doped graphene nanoribbons. *Physical Review B*, 78(15):155427, 2008.

[106] S. Grimme. Semiempirical gga-type density functional constructed with a long-range dispersion correction. *Journal of Computational Chemistry*, 27(15):1787–1799, 2006.

[107] A. Grüneis, C. Attaccalite, T. Pichler, V. Zabolotnyy, H. Shiozawa, S. Molodtsov, D. Inosov, A. Koitzsch, M. Knupfer, and J. Schiessling. Electron-electron correlation in graphite: a combined angle-resolved photoemission and first-principles study. *Physical Review Letters*, 100(3):037601, 2008.

[108] A. Grüneis and D. V. Vyalikh. Tunable hybridization between electronic states of graphene and a metal surface. *Physical Review B*, 77(19):193401, 2008.

[109] F. Guinea, A. C. Neto, and N. Peres. Electronic states and landau levels in graphene stacks. *Physical Review B*, 73(24):245426, 2006.

[110] O. Gülseren, T. Yildirim, and S. Ciraci. Systematic ab initio study of curvature effects in carbon nanotubes. *Physical Review B*, 65(15):153405, 2002.

[111] Y. Guo, W. Guo, and C. Chen. Semiconducting to half-metallic to metallic transition on spin-resolved zigzag bilayer graphene nanoribbons. *The Journal of Physical Chemistry C*, 114(30):13098–13105, 2010.

[112] V. Gusynin, S. Sharapov, and J. Carbotte. Magneto-optical conductivity in graphene. *Journal of Physics: Condensed Matter*, 19(2):026222, 2006.

[113] V. Gusynin, S. Sharapov, and J. Carbotte. Ac conductivity of graphene: from tight-binding model to 2+ 1-dimensional quantum electrodynamics. *International Journal of Modern Physics B*, 21(27):4611–4658, 2007.

[114] M. Gyamfi, T. Eelbo, M. Waśniowska, and R. Wiesendanger. Fe adatoms on graphene/ru (0001): Adsorption site and local electronic properties. *Physical Review B*, 84(11):113403, 2011.

[115] R. Haddon. Electronic structure, conductivity and superconductivity of alkali metal doped (c60). *Accounts of Chemical Research*, 25(3):127–133, 1992.

[116] N. Hamada, S.-i. Sawada, and A. Oshiyama. New one-dimensional conductors: graphitic microtubules. *Physical Review Letters*, 68(10):1579, 1992.

[117] M. Y. Han, B. Özyilmaz, Y. Zhang, and P. Kim. Energy band-gap engineering of graphene nanoribbons. *Physical Review Letters*, 98(20):206805, 2007.

[118] W. Han, R. K. Kawakami, M. Gmitra, and J. Fabian. Graphene spintronics. *Nature Nanotechnology*, 9(10):794–807, 2014.

[119] X. He, F. Léonard, and J. Kono. Uncooled carbon nanotube photodetectors. *Advanced Optical Materials*, 3(8):989–1011, 2015.

[120] Z. He, K. He, A. W. Robertson, A. I. Kirkland, D. Kim, J. Ihm, E. Yoon, G.-D. Lee, and J. H. Warner. Atomic structure and dynamics of metal dopant pairs in graphene. *Nano Letters*, 14(7):3766–3772, 2014.

[121] T. Hikihara, X. Hu, H.-H. Lin, and C.-Y. Mou. Ground-state properties of nanographite systems with zigzag edges. *Physical Review B*, 68(3):035432, 2003.

[122] T. Hirahara, N. Fukui, T. Shirasawa, M. Yamada, M. Aitani, H. Miyazaki, M. Matsunami, S. Kimura, T. Takahashi, and S. Hasegawa. Atomic and electronic structure of ultrathin bi (111) films grown on bi 2 te 3 (111) substrates: evidence for a strain-induced topological phase transition. *Physical Review Letters*, 109(22):227401, 2012.

[123] T. Hirahara, T. Nagao, I. Matsuda, G. Bihlmayer, E. Chulkov, Y. M. Koroteev, P. Echenique, M. Saito, and S. Hasegawa. Role of spin-orbit coupling and hybridization effects in the electronic structure of ultrathin bi films. *Physical Review Letters*, 97(14):146803, 2006.

[124] T. Hirahara, T. Shirai, T. Hajiri, M. Matsunami, K. Tanaka, S. Kimura, S. Hasegawa, and K. Kobayashi. Role of quantum and surface-state effects in the bulk fermi-level position of ultrathin bi films. *Physical Review Letters*, 115(10):106803, 2015.

[125] J.-H. Ho, C. Lu, C. Hwang, C. Chang, and M. F. Lin. Coulomb excitations in aa-and ab-stacked bilayer graphites. *Physical Review B*, 74(8):085406, 2006.

[126] Y.-H. Ho, Y.-H. Chiu, D.-H. Lin, C.-P. Chang, and M. F. Lin. Magneto-optical selection rules in bilayer bernal graphene. *ACS Nano*, 4(3):1465–1472, 2010.

[127] P. Hofmann. The surfaces of bismuth: Structural and electronic properties. *Progress in Surface Science*, 81(5):191–245, 2006.

[128] C. Howard, M. Dean, and F. Withers. Phonons in potassium-doped graphene: the effects of electron-phonon interactions, dimensionality, and adatom ordering. *Physical Review B*, 84(24):241404, 2011.

[129] C.-H. Hsu, V. Ozolins, and F.-C. Chuang. First-principles study of bi and sb intercalated graphene on sic (0001) substrate. *Surface Science*, 616:149–154, 2013.

[130] P.-J. Hsu, A. Kubetzka, A. Finco, N. Romming, K. von Bergmann, and R. Wiesendanger. Electric-field-driven switching of individual magnetic skyrmions. *Nature Nanotechnology*, 12(2):123, 2017.

[131] X. Hu, H. Tian, W. Zheng, S. Yu, L. Qiao, C. Qu, and Q. Jiang. Metallic-semiconducting phase transition of the edge-oxygenated armchair graphene nanoribbons. *Chemical Physics Letters*, 501(1):64–67, 2010.

[132] B. Huang, Z. Li, Z. Liu, G. Zhou, S. Hao, J. Wu, B.-L. Gu, and W. Duan. Adsorption of gas molecules on graphene nanoribbons and its implication for nanoscale molecule sensor. *The Journal of Physical Chemistry C*, 112(35):13442–13446, 2008.

[133] H. Huang, Z. Li, H. Kreuzer, and W. Wang. Disintegration of graphene nanoribbons in large electrostatic fields. *Physical Chemistry Chemical Physics*, 16(30):15927–15933, 2014.

[134] H. Huang, D. Wei, J. Sun, S. L. Wong, Y. P. Feng, A. C. Neto, and A. T. S. Wee. Spatially resolved electronic structures of atomically precise armchair graphene nanoribbons. *Scientific Reports*, 2:983, 2012.

[135] J. Y. Huang, L. Qi, and J. Li. In situ imaging of layer-by-layer sublimation of suspended graphene. *Nano Research*, 3(1):43–50, 2010.

[136] Y. Huang, C. Chang, and M.-F. Lin. Magnetic and quantum confinement effects on electronic and optical properties of graphene ribbons. *Nanotechnology*, 18(49):495401, 2007.

[137] Y.-K. Huang, S.-C. Chen, Y.-H. Ho, C.-Y. Lin, and M. F. Lin. Feature-rich magnetic quantization in sliding bilayer graphenes. *Scientific Reports*, 4:7509, 2014.

[138] S. Hüfner. *Photoelectron Spectroscopy: Principles and Applications.* Springer Science & Business Media, 2013.

[139] S. Iijima. Helical microtubules of graphitic carbon. *Nature*, 354(6348):56, 1991.

[140] S. Iijima and T. Ichihashi. Single-shell carbon nanotubes of 1-nm diameter. *Nature*, 363(6430):603–605, 1993.

[141] J. D. Jackson. *Classical Electrodynamics, 3rd Ed.* Wiley, London, 1998.

[142] W.-L. W. D.-X. Y. Jason Lee, Wen-Chuan Tian. Two-dimensional pnictogen honeycomb lattice: Structure, on-site spin-orbit coupling and spin polarization. *Scientific Reports*, 5(11512), 2015.

[143] I.-Y. Jeon, M. J. Ju, J. Xu, H.-J. Choi, J.-M. Seo, I. T. Choi, H. M. Kim, J. C. Kim, and J.-J. Lee. Fluorine: Edge-fluorinated graphene nanoplatelets as high performance electrodes for dye-sensitized solar cells and lithium ion batteries (adv. funct. mater. 8/2015). *Adv. Funct. Mater.*, 25(8):1328–1328, 2015.

[144] I.-Y. Jeon, M. J. Ju, J. Xu, H.-J. Choi, J.-M. Seo, M.-J. Kim, I. T. Choi, H. M. Kim, J. C. Kim, and J.-J. Lee. Edge-fluorinated graphene nanoplatelets as high performance electrodes for dye-sensitized solar cells and lithium ion batteries. *Advanced Functional Materials*, 25(8):1170–1179, 2015.

[145] L. Jiao, X. Wang, G. Diankov, H. Wang, and H. Dai. Facile synthesis of high-quality graphene nanoribbons. *Nature Nanotechnology*, 5(5):321–325, 2010.

[146] L. Jiao, L. Zhang, L. Ding, J. Liu, and H. Dai. Aligned graphene nanoribbons and crossbars from unzipped carbon nanotubes. *Nano Research*, 3(6):387–394, 2010.

[147] L. Jiao, L. Zhang, X. Wang, G. Diankov, and H. Dai. Narrow graphene nanoribbons from carbon nanotubes. *Nature*, 458(7240):877–880, 2009.

[148] J. L. Johnson, A. Behnam, S. Pearton, and A. Ural. Hydrogen sensing using pd-functionalized multi-layer graphene nanoribbon networks. *Advanced Materials*, 22(43):4877–4880, 2010.

[149] O. Jost, A. Gorbunov, W. Pompe, T. Pichler, R. Friedlein, M. Knupfer, M. Reibold, H.-D. Bauer, L. Dunsch, and M. Golden. Diameter grouping in bulk samples of single-walled carbon nanotubes from optical absorption spectroscopy. *Applied Physics Letters*, 75(15):2217–2219, 1999.

[150] C. L. Kane and E. Mele. Size, shape, and low energy electronic structure of carbon nanotubes. *Physical Review Letters*, 78(10):1932, 1997.

[151] C. L. Kane and E. J. Mele. Quantum spin hall effect in graphene. *Physical Review Letters*, 95(22):226801, 2005.

[152] J. Kang, F. Wu, S.-S. Li, J.-B. Xia, and J. Li. Antiferromagnetic coupling and spin filtering in asymmetrically hydrogenated graphene nanoribbon homojunction. *Appl. Phys. Lett.*, 100(15):153102, 2012.

[153] H. Kataura, Y. Kumazawa, Y. Maniwa, I. Umezu, S. Suzuki, Y. Ohtsuka, and Y. Achiba. Optical properties of single-wall carbon nanotubes. *Synthetic Metals*, 103(1-3):2555–2558, 1999.

[154] M. Katkov, V. Sysoev, A. Gusel'Nikov, I. Asanov, L. Bulusheva, and A. Okotrub. A backside fluorine-functionalized graphene layer for ammonia detection. *Phys. Chem. Chem. Phys.*, 17(1):444–450, 2015.

[155] K. Kim, A. Sussman, and A. Zettl. Graphene nanoribbons obtained by electrically unwrapping carbon nanotubes. *ACS Nano*, 4(3):1362–1366, 2010.

[156] W. Y. Kim and K. S. Kim. Prediction of very large values of magnetoresistance in a graphene nanoribbon device. *Nature Nanotechnol.*, 3(7):408, 2008.

[157] Y.-H. Kim, H.-S. Sim, and K. Chang. Electronic structure of collapsed c, bn, and bc 3 nanotubes. *Current Applied Physics*, 1(1):39–44, 2001.

[158] C. Kisielowski, B. Freitag, M. Bischoff, H. Van Lin, S. Lazar, G. Knippels, P. Tiemeijer, M. van der Stam, S. von Harrach, and M. Stekelenburg. Detection of single atoms and buried defects in three dimensions by aberration-corrected electron microscope with 0.5-å information limit. *Microscopy and Microanalysis*, 14(5):469–477, 2008.

[159] Z. Klusek. Investigations of splitting of the $\pi$ bands in graphite by scanning tunneling spectroscopy. *Applied Surface Science*, 151(3):251–261, 1999.

[160] Y. Kobayashi, K.-i. Fukui, T. Enoki, K. Kusakabe, and Y. Kaburagi. Observation of zigzag and armchair edges of graphite using scanning tunneling microscopy and spectroscopy. *Physical Review B*, 71(19):193406, 2005.

[161] M. Koch, F. Ample, C. Joachim, and L. Grill. Voltage-dependent conductance of a single graphene nanoribbon. *Nature Nanotechnology*, 7(11):713, 2012.

[162] M. Kociak, L. Henrard, O. Stéphan, K. Suenaga, and C. Colliex. Plasmons in layered nanospheres and nanotubes investigated by spatially resolved electron energy-loss spectroscopy. *Physical Review B*, 61(20):13936, 2000.

[163] W. Kohn and L. J. Sham. Self-consistent equations including exchange and correlation effects. *Physical Review*, 140(4A):A1133, 1965.

[164] T. Kondo, S. Casolo, T. Suzuki, T. Shikano, M. Sakurai, Y. Harada, M. Saito, M. Oshima, M. I. Trioni, and G. F. Tantardini. Atomic-scale characterization of nitrogen-doped graphite: Effects of dopant nitrogen on the local electronic structure of the surrounding carbon atoms. *Physical Review B*, 86(3):035436, 2012.

[165] M. Koshino. Stacking-dependent optical absorption in multilayer graphene. *New Journal of Physics*, 15(1):015010, 2013.

[166] D. V. Kosynkin, A. L. Higginbotham, A. Sinitskii, J. R. Lomeda, A. Dimiev, B. K. Price, and J. M. Tour. Longitudinal unzipping of carbon nanotubes to form graphene nanoribbons. *Nature*, 458(7240):872–876, 2009.

[167] D. V. Kosynkin, W. Lu, A. Sinitskii, G. Pera, Z. Sun, and J. M. Tour. Highly conductive graphene nanoribbons by longitudinal splitting of carbon nanotubes using potassium vapor. *ACS Nano*, 5(2):968–974, 2011.

[168] G. Kresse and J. Furthmüller. Efficient iterative schemes for ab initio total-energy calculations using a plane-wave basis set. *Physical Review B*, 54(16):11169, 1996.

[169] H. W. Kroto. C 60 buckminsterfullerene. *MRS Online Proceedings Library Archive*, 206, 1990.

[170] H. W. Kroto, J. R. Heath, S. C. O'Brien, R. F. Curl, and R. E. Smalley. C 60: buckminsterfullerene. *Nature*, 318(6042):162–163, 1985.

[171] P. Kumar, L. Panchakarla, and C. Rao. Laser-induced unzipping of carbon nanotubes to yield graphene nanoribbons. *Nanoscale*, 3(5):2127–2129, 2011.

[172] L. Lai, J. Lu, L. Wang, G. Luo, J. Zhou, R. Qin, Y. Chen, H. Li, Z. Gao, and G. Li. Magnetism in carbon nanoscrolls: Quasi-half-metals and half-metals in pristine hydrocarbons. *Nano Research*, 2(11):844–850, 2009.

[173] Y. Lai, J. Ho, C. Chang, and M. F. Lin. Magnetoelectronic properties of bilayer bernal graphene. *Physical Review B*, 77(8):085426, 2008.

[174] P. E. Lammert, P. Zhang, and V. H. Crespi. Gapping by squashing: Metal-insulator and insulator-metal transitions in collapsed carbon nanotubes. *Physical Review Letters*, 84(11):2453, 2000.

[175] P. Lauffer, K. Emtsev, R. Graupner, T. Seyller, L. Ley, S. Reshanov, and H. Weber. Atomic and electronic structure of few-layer graphene on sic (0001) studied with scanning tunneling microscopy and spectroscopy. *Physical Review B*, 77(15):155426, 2008.

[176] J. G. Lavin, S. Subramoney, R. S. Ruoff, S. Berber, and D. Tomanek. Scrolls and nested tubes in multiwall carbon nanotubes. *Carbon*, 40(7):1123–1130, 2002.

[177] N. B. Le and L. M. Woods. Folded graphene nanoribbons with single and double closed edges. *Physical Review B*, 85(3):035403, 2012.

[178] C. Lee, S. Chen, C. Yang, W. Su, and M. Lin. Low-energy electronic structures of nanotube–nanoribbon hybrid systems. *Computer Physics Communications*, 182(1):68–70, 2011.

[179] C.-H. Lee, C.-K. Yang, M. F. Lin, C.-P. Chang, and W.-S. Su. Structural and electronic properties of graphene nanotube–nanoribbon hybrids. *Physical Chemistry Chemical Physics*, 13(9):3925–3931, 2011.

[180] G. Lee and K. Cho. Electronic structures of zigzag graphene nanoribbons with edge hydrogenation and oxidation. *Physical Review B*, 79(16):165440, 2009.

[181] J.-K. Lee, S.-C. Lee, J.-P. Ahn, S.-C. Kim, J. I. Wilson, and P. John. The growth of aa graphite on (111) diamond. *The Journal of Chemical Physics*, 129(23):234709, 2008.

[182] R. Lee, H. Kim, J. Fischer, A. Thess, and R. E. Smalley. Conductivity enhancement in single-walled carbon nanotube bundles doped with k and br. *Nature*, 388(6639):255, 1997.

[183] M. Lentzen. Progress in aberration-corrected high-resolution transmission electron microscopy using hardware aberration correction. *Microscopy and Microanalysis*, 12(3):191–205, 2006.

[184] G. Levy, W. Nettke, B. Ludbrook, C. Veenstra, and A. Damascelli. Deconstruction of resolution effects in angle-resolved photoemission. *Physical Review B*, 90(4):045150, 2014.

[185] B. Li, L. Zhou, D. Wu, H. Peng, K. Yan, Y. Zhou, and Z. Liu. Photochemical chlorination of graphene. *ACS Nano*, 5(7):5957–5961, 2011.

[186] C. Li and T.-W. Chou. A structural mechanics approach for the analysis of carbon nanotubes. *International Journal of Solids and Structures*, 40(10):2487–2499, 2003.

[187] G. Li, A. Luican, and E. Y. Andrei. Scanning tunneling spectroscopy of graphene on graphite. *Physical Review Letters*, 102(17):176804, 2009.

[188] G. Li, A. Luican, J. L. Dos Santos, A. C. Neto, A. Reina, J. Kong, and E. Andrei. Observation of van hove singularities in twisted graphene layers. *Nature Physics*, 6(2):109–113, 2010.

[189] J. Li, Q. Peng, G. Bai, and W. Jiang. Carbon scrolls produced by high energy ball milling of graphite. *Carbon*, 43(13):2817–2833, 2005.

[190] L. Li, J. G. Checkelsky, Y. S. Hor, C. Uher, A. F. Hebard, R. J. Cava, and N. Ong. Phase transitions of dirac electrons in bismuth. *Science*, 321(5888):547–550, 2008.

[191] N. Li, Z. Chen, W. Ren, F. Li, and H.-M. Cheng. Flexible graphene-based lithium ion batteries with ultrafast charge and discharge rates. *Proceedings of the National Academy of Sciences*, 109(43):17360–17365, 2012.

[192] X. Li, J. T. Mullen, Z. Jin, K. M. Borysenko, M. B. Nardelli, and K. W. Kim. Intrinsic electrical transport properties of monolayer silicene and mos 2 from first principles. *Physical Review B*, 87(11):115418, 2013.

[193] X. Li, X. Wang, L. Zhang, S. Lee, and H. Dai. Chemically derived, ultra-smooth graphene nanoribbon semiconductors. *Science*, 319(5867):1229–1232, 2008.

[194] Y. Li, H. Chen, L. Y. Voo, J. Ji, G. Zhang, G. Zhang, F. Zhang, and X. Fan. Synthesis of partially hydrogenated graphene and brominated graphene. *Journal of Materials Chemistry*, 22(30):15021–15024, 2012.

[195] Y. Li, M. A. Trujillo, E. Fu, B. Patterson, L. Fei, Y. Xu, S. Deng, S. Smirnov, and H. Luo. Bismuth oxide: a new lithium-ion battery anode. *Journal of Materials Chemistry A*, 1(39):12123–12127, 2013.

[196] Y. Li, Z. Zhou, S. Zhang, and Z. Chen. Mos2 nanoribbons: high stability and unusual electronic and magnetic properties. *Journal of the American Chemical Society*, 130(49):16739–16744, 2008.

[197] Z. Li, H. Qian, J. Wu, B.-L. Gu, and W. Duan. Role of symmetry in the transport properties of graphene nanoribbons under bias. *Physical Review Letters*, 100(20):206802, 2008.

[198] C. Lian and J. Ni. Strain induced phase transitions in silicene bilayers: a first principles and tight-binding study. *AIP Advances*, 3(5):052102, 2013.

[199] L. Liang, J. Wang, W. Lin, B. G. Sumpter, V. Meunier, and M. Pan. Electronic bandgap and edge reconstruction in phosphorene materials. *Nano Letters*, 14(11):6400–6406, 2014.

[200] S. Liang, Z. Ma, G. Wu, N. Wei, L. Huang, H. Huang, H. Liu, S. Wang, and L.-M. Peng. Microcavity-integrated carbon nanotube photodetectors. *ACS Nano*, 10(7):6963–6971, 2016.

[201] M. P. Lima, A. Fazzio, and A. J. da Silva. Edge effects in bilayer graphene nanoribbons: Ab initio total-energy density functional theory calculations. *Physical Review B*, 79(15):153401, 2009.

[202] C. Y. Lin, T. N. Do, Y. K. Huang, and M. F. Lin. *Optical Properties of Graphene in Magnetic and Electric Fields*. IOP Publishing, e-book, 2017.

[203] C.-Y. Lin, J.-Y. Wu, Y.-J. Ou, Y.-H. Chiu, and M. F. Lin. Magnetoelectronic properties of multilayer graphenes. *Physical Chemistry Chemical Physics*, 17(39):26008–26035, 2015.

[204] J. Lin, Z. Peng, C. Xiang, G. Ruan, Z. Yan, D. Natelson, and J. M. Tour. Graphene nanoribbon and nanostructured sno2 composite anodes for lithium ion batteries. *ACS Nano*, 7(7):6001–6006, 2013.

[205] J. Lin, A.-R. O. Raji, K. Nan, Z. Peng, Z. Yan, E. L. Samuel, D. Natelson, and J. M. Tour. Iron oxide nanoparticle and graphene nanoribbon composite as an anode material for high-performance li-ion batteries. *Advanced Functional Materials*, 24(14):2044–2048, 2014.

[206] M.-C. Lin, M. Gong, B. Lu, Y. Wu, D.-Y. Wang, M. Guan, M. Angell, C. Chen, J. Yang, and B.-J. Hwang. An ultrafast rechargeable aluminium-ion battery. *Nature*, 520(7547):324, 2015.

[207] M. F. Lin and D.-S. Chuu. Persistent currents in toroidal carbon nanotubes. *Physical Review B*, 57(11):6731, 1998.

[208] M. F. Lin, D.-S. Chuu, C.-S. Huang, Y.-K. Lin, and K. W.-K. Shung. Collective excitations in a single-layer carbon nanotube. *Physical Review B*, 53(23):15493, 1996.

[209] M. F. Lin and K. W.-K. Shung. Elementary excitations in cylindrical tubules. *Physical Review B*, 47(11):6617, 1993.

[210] M. F. Lin and K. W.-K. Shung. Plasmons and optical properties of carbon nanotubes. *Physical Review B*, 50(23):17744, 1994.

[211] M. F. Lin and K. W.-K. Shung. Magnetization of graphene tubules. *Physical Review B*, 52(11):8423, 1995.

[212] M. F. Lin and K. W.-K. Shung. Magnetoconductance of carbon nanotubes. *Physical Review B*, 51(12):7592, 1995.

[213] S.-Y. Lin, S.-L. Chang, H.-H. Chen, S.-H. Su, J.-C. Huang, and M.-F. Lin. Substrate-induced structures of bismuth adsorption on graphene: a first principles study. *Physical Chemistry Chemical Physics*, 18(28):18978–18984, 2016.

[214] S.-Y. Lin, S.-L. Chang, F.-L. Shyu, J.-M. Lu, and M.-F. Lin. Feature-rich electronic properties in graphene ripples. *Carbon*, 86:207–216, 2015.

[215] S.-Y. Lin, S.-L. Chang, N. T. T. Tran, P.-H. Yang, and M.-F. Lin. H–si bonding-induced unusual electronic properties of silicene: a method to identify hydrogen concentration. *Physical Chemistry Chemical Physics*, 17(39):26443–26450, 2015.

[216] S.-Y. Lin, Y.-T. Lin, N. T. T. Tran, W.-P. Su, and M.-F. Lin. Feature-rich electronic properties of aluminum-adsorbed graphenes. *Carbon*, 120:209–218, 2017.

[217] Y.-T. Lin, H.-C. Chung, P.-H. Yang, S.-Y. Lin, and M. F. Lin. Adatom bond-induced geometric and electronic properties of passivated armchair graphene nanoribbons. *Physical Chemistry Chemical Physics*, 17(25):16545–16552, 2015.

[218] Y.-T. Lin, S.-Y. Lin, Y.-H. Chiu, and M.-F. Lin. Alkali-created rich properties in grapheme nanoribbons: Chemical bondings. *Scientific Reports*, 7, 2017.

[219] F. Liu, S. Song, D. Xue, and H. Zhang. Folded structured graphene paper for high performance electrode materials. *Advanced Materials*, 24(8):1089–1094, 2012.

[220] J. Liu, H. Dai, J. H. Hafner, and D. T. Colbert. Fullerene 'crop circles'. *Nature*, 385(6619):780, 1997.

[221] X. H. Liu, J. W. Wang, Y. Liu, H. Zheng, A. Kushima, S. Huang, T. Zhu, S. X. Mao, J. Li, and S. Zhang. In situ transmission electron microscopy of electrochemical lithiation, delithiation and deformation of individual graphene nanoribbons. *Carbon*, 50(10):3836–3844, 2012.

[222] Y. Liu and R. E. Allen. Electronic structure of the semimetals bi and sb. *Phys. Rev. B*, 52:1566–1577, Jul 1995.

[223] Y. Liu, X. Dong, and P. Chen. Biological and chemical sensors based on graphene materials. *Chemical Society Reviews*, 41(6):2283–2307, 2012.

[224] Z. Liu, K. Suenaga, P. J. Harris, and S. Iijima. Open and closed edges of graphene layers. *Physical Review Letters*, 102(1):015501, 2009.

[225] Z. Liu, J. Yang, F. Grey, J. Z. Liu, Y. Liu, Y. Wang, Y. Yang, Y. Cheng, and Q. Zheng. Observation of microscale superlubricity in graphite. *Physical Review Letters*, 108(20):205503, 2012.

[226] P. Longe and S. Bose. Collective excitations in metallic graphene tubules. *Physical Review B*, 48(24):18239, 1993.

[227] C. Lu, C.-P. Chang, Y.-C. Huang, R.-B. Chen, and M. Lin. Influence of an electric field on the optical properties of few-layer graphene with ab stacking. *Physical Review B*, 73(14):144427, 2006.

[228] D.-b. Lu, Y.-l. Song, Z.-x. Yang, and G.-q. Li. Energy gap modulation of graphene nanoribbons by f termination. *Applied Surface Science*, 257(15):6440–6444, 2011.

[229] J. P. Lu. Novel magnetic properties of carbon nanotubes. *Physical Review Letters*, 74(7):1123, 1995.

[230] W. Lu, G. Ruan, B. Genorio, Y. Zhu, B. Novosel, Z. Peng, and J. M. Tour. Functionalized graphene nanoribbons via anionic polymerization initiated by alkali metal-intercalated carbon nanotubes. *ACS Nano*, 7(3):2669–2675, 2013.

[231] A. Luican, G. Li, A. Reina, J. Kong, R. Nair, K. S. Novoselov, A. K. Geim, and E. Andrei. Single-layer behavior and its breakdown in twisted graphene layers. *Physical Review Letters*, 106(12):126802, 2011.

[232] V. Lukose, R. Shankar, and G. Baskaran. Novel electric field effects on landau levels in graphene. *Physical Review Letters*, 98(11):116802, 2007.

[233] G. Z. Magda, X. Jin, I. Hagymási, P. Vancsó, Z. Osváth, P. Nemes-Incze, C. Hwang, L. P. Biro, and L. Tapaszto. Room-temperature magnetic order on zigzag edges of narrow graphene nanoribbons. *Nature*, 514(7524):608–611, 2014.

[234] Y. Mao, W. Hao, X. Wei, J. Yuan, and J. Zhong. Edge-adsorption of potassium adatoms on graphene nanoribbon: A first principle study. *Applied Surface Science*, 280:698–704, 2013.

[235] R. Martel, T. Schmidt, H. Shea, T. Hertel, and P. Avouris. Single- and multi-wall carbon nanotube field-effect transistors. *Applied Physics Letters*, 73(17):2447–2449, 1998.

[236] T. Mashoff, M. Takamura, S. Tanabe, H. Hibino, F. Beltram, and S. Heun. Hydrogen storage with titanium-functionalized graphene. *Applied Physics Letters*, 103(1):013903, 2013.

[237] S. Masubuchi, M. Ono, K. Yoshida, K. Hirakawa, and T. Machida. Fabrication of graphene nanoribbon by local anodic oxidation lithography using atomic force microscope. *Applied Physics Letters*, 94(8):082107, 2009.

[238] M. J. McAllister, J.-L. Li, D. H. Adamson, H. C. Schniepp, A. A. Abdala, J. Liu, M. Herrera-Alonso, D. L. Milius, R. Car, and R. K. Prud'homme. Single sheet functionalized graphene by oxidation and thermal expansion of graphite. *Chemistry of Materials*, 19(18):4396–4404, 2007.

[239] E. McCann. Asymmetry gap in the electronic band structure of bilayer graphene. *Physical Review B*, 74(16):161403, 2006.

[240] P. V. Medeiros, A. J. Mascarenhas, F. de Brito Mota, and C. M. C. de Castilho. A dft study of halogen atoms adsorbed on graphene layers. *Nanotech.*, 21(48):485701, 2010.

[241] L. Meng, W.-Y. He, H. Zheng, M. Liu, H. Yan, W. Yan, Z.-D. Chu, K. Bai, R.-F. Dou, and Y. Zhang. Strain-induced one-dimensional landau level quantization in corrugated graphene. *Physical Review B*, 87(20):205405, 2013.

[242] J. Mintmire, B. Dunlap, and C. White. Are fullerene tubules metallic? *Physical Review Letters*, 68(5):631, 1992.

[243] J. Misewich, R. Martel, P. Avouris, J. Tsang, S. Heinze, and J. Tersoff. Electrically induced optical emission from a carbon nanotube fet. *Science*, 300(5620):783–786, 2003.

[244] J. Mittal and K. Lin. Carbon nanotube-based interconnections. *Journal of Materials Science*, 52(2):643–662, 2017.

[245] T. Miyake and S. Saito. Quasiparticle band structure of carbon nanotubes. *Physical Review B*, 68(15):155424, 2003.

[246] G. Mpourmpakis, E. Tylianakis, and G. E. Froudakis. Carbon nanoscrolls: a promising material for hydrogen storage. *Nano Letters*, 7(7):1893–1897, 2007.

[247] T. Mueller, M. Kinoshita, M. Steiner, V. Perebeinos, A. A. Bol, D. B. Farmer, and P. Avouris. Efficient narrow-band light emission from a single carbon nanotube p–n diode. *Nature Nanotechnology*, 5(1):27–31, 2010.

[248] M. Myodo, M. Inaba, , K. Ohara, R. Kato, M. Kobayashi, Y. Hirano, K. Suzuki, and H. Kawarada. Large-current-controllable carbon nanotube field-effect transistor in electrolyte solution. *Applied Physics Letters*, 106(21):213503, 2015.

[249] K. Nakada, M. Fujita, G. Dresselhaus, and M. S. Dresselhaus. Edge state in graphene ribbons: Nanometer size effect and edge shape dependence. *Physical Review B*, 54(24):17954, 1996.

[250] A. G. Nasibulin, P. V. Pikhitsa, H. Jiang, D. P. Brown, A. V. Krasheninnikov, A. S. Anisimov, P. Queipo, A. Moisala, D. Gonzalez, and G. Lientschnig. A novel hybrid carbon material. *Nature Nanotechnology*, 2(3):156–161, 2007.

[251] A. C. Neto, F. Guinea, N. M. Peres, K. S. Novoselov, and A. K. Geim. The electronic properties of graphene. *Reviews of Modern Physics*, 81(1):109, 2009.

[252] T. Ngoc Thanh Thuy, N. Duy Khanh, G. Olga E, and L. Ming-Fa. Coverage-dependent essential properties of halogenated graphene: A dft study. *Scientific Reports*, 7(17858):1–13, 2017.

[253] D. K. Nguyen, Y.-T. Lin, S.-Y. Lin, Y.-H. Chiu, N. T. T. Tran, and M. Fa-Lin. Fluorination-enriched electronic and magnetic properties in graphene nanoribbons. *Physical Chemistry Chemical Physics*, 19(31):20667–20676, 2017.

[254] Y. Niimi, T. Matsui, H. Kambara, K. Tagami, M. Tsukada, and H. Fukuyama. Scanning tunneling microscopy and spectroscopy of the electronic local density of states of graphite surfaces near monoatomic step edges. *Physical Review B*, 73(8):085421, 2006.

[255] K. S. Novoselov, A. K. Geim, S. V. Morozov, D. Jiang, Y. Zhang, S. V. Dubonos, I. V. Grigorieva, and A. A. Firsov. Electric field effect in atomically thin carbon films. *Science*, 306(5696):666–669, 2004.

[256] T. W. Odom, J.-L. Huang, P. Kim, and C. M. Lieber. Atomic structure and electronic properties of single-walled carbon nanotubes. *Nature*, 391(6662):62–64, 1998.

[257] T. Ohta, A. Bostwick, J. L. McChesney, T. Seyller, K. Horn, and E. Rotenberg. Interlayer interaction and electronic screening in multilayer graphene investigated with angle-resolved photoemission spectroscopy. *Physical Review Letters*, 98(20):206802, 2007.

[258] T. Ohta, A. Bostwick, T. Seyller, K. Horn, and E. Rotenberg. Controlling the electronic structure of bilayer graphene. *Science*, 313(5789):951–954, 2006.

[259] C. Ophus, H. I. Rasool, M. Linck, A. Zettl, and J. Ciston. Automatic software correction of residual aberrations in reconstructed hrtem exit waves of crystalline samples. *Advanced Structural and Chemical Imaging*, 2(1):15, 2017.

[260] M. Ouyang, J.-L. Huang, C. L. Cheung, and C. M. Lieber. Energy gaps in" metallic" single-walled carbon nanotubes. *Science*, 292(5517):702–705, 2001.

[261] M. C. Paiva, W. Xu, M. Fernanda Proença, R. M. Novais, E. Lægsgaard, and F. Besenbacher. Unzipping of functionalized multiwall carbon nanotubes induced by stm. *Nano Letters*, 10(5):1764–1768, 2010.

[262] D. Pandey, R. Reifenberger, and R. Piner. Scanning probe microscopy study of exfoliated oxidized graphene sheets. *Surface Science*, 602(9):1607–1613, 2008.

[263] M. Papagno, S. Rusponi, P. M. Sheverdyaeva, S. Vlaic, M. Etzkorn, D. Pacilé, P. Moras, C. Carbone, and H. Brune. Large band gap opening between graphene dirac cones induced by na adsorption onto an ir superlattice. *ACS Nano*, 6(1):199–204, 2011.

[264] U. K. Parashar, S. Bhandari, R. K. Srivastava, D. Jariwala, and A. Srivastava. Single step synthesis of graphene nanoribbons by catalyst particle size dependent cutting of multiwalled carbon nanotubes. *Nanoscale*, 3(9):3876–3882, 2011.

[265] C. Park, J. Ryou, S. Hong, B. G. Sumpter, G. Kim, and M. Yoon. Electronic properties of bilayer graphene strongly coupled to interlayer stacking and an external electric field. *Physical Review Letters*, 115(1):015502, 2015.

[266] K. K. Paulla and A. A. Farajian. Stacking stability, emergence of magnetization and electromechanical nanosensing in bilayer graphene nanoribbons. *Journal of Physics: Condensed Matter*, 25(11):115303, 2013.

[267] S. Pekker, A. Janossy, L. Mihaly, O. Chauvet, M. Carrard, and L. Forró. Single-crystalline (kc60) n: a conducting linear alkali fulleride polymer. *Science*, 265(5175):1077–1078, 1994.

[268] X. Peng and S. Velasquez. Strain modulated band gap of edge passivated armchair graphene nanoribbons. *Applied Physics Letters*, 98(2):023112, 2011.

[269] G. Pennington and N. Goldsman. Semiclassical transport and phonon scattering of electrons in semiconducting carbon nanotubes. *Physical Review B*, 68(4):045426, 2003.

[270] J. P. Perdew, K. Burke, and M. Ernzerhof. Generalized gradient approximation made simple. *Physical Review Letters*, 77(18):3865, 1996.

[271] K. Pi, K. McCreary, W. Bao, W. Han, Y. Chiang, Y. Li, S.-W. Tsai, C. Lau, and R. Kawakami. Electronic doping and scattering by transition metals on graphene. *Physical Review B*, 80(7):075406, 2009.

[272] T. Pichler, M. Knupfer, M. Golden, J. Fink, A. Rinzler, and R. Smalley. Localized and delocalized electronic states in single-wall carbon nanotubes. *Physical Review Letters*, 80(21):4729, 1998.

[273] D. T. Pierce. Spin-polarized electron microscopy. *Physica Scripta*, 38(2):291, 1988.

[274] D. Pierucci, H. Sediri, M. Hajlaoui, J.-C. Girard, T. Brumme, M. Calandra, E. Velez-Fort, G. Patriarche, M. G. Silly, and G. Ferro. Evidence for flat bands near the fermi level in epitaxial rhombohedral multilayer graphene. *ACS Nano*, 9(5):5432–5439, 2015.

[275] O. Pietzsch, A. Kubetzka, M. Bode, and R. Wiesendanger. Real-space observation of dipolar antiferromagnetism in magnetic nanowires by spin-polarized scanning tunneling spectroscopy. *Physical Review Letters*, 84(22):5212, 2000.

[276] O. Pietzsch, A. Kubetzka, M. Bode, and R. Wiesendanger. Spin-polarized scanning tunneling spectroscopy of nanoscale cobalt islands on cu (111). *Physical Review Letters*, 92(5):057202, 2004.

[277] H. L. Poh, P. Šimek, Z. Sofer, and M. Pumera. Halogenation of graphene with chlorine, bromine, or iodine by exfoliation in a halogen atmosphere. *Chemistry-A European Journal*, 19(8):2655–2662, 2013.

[278] E. Prada, P. San-Jose, and L. Brey. Zero landau level in folded graphene nanoribbons. *Physical Review Letters*, 105(10):106802, 2010.

[279] M. Pumera. Graphene-based nanomaterials for energy storage. *Energy & Environmental Science*, 4(3):668–674, 2011.

[280] F. Pyatkov, V. Fütterling, S. Khasminskaya, B. S. Flavel, F. Hennrich, M. M. Kappes, R. Krupke, and W. H. Pernice. Cavity-enhanced light emission from electrically driven carbon nanotubes. *Nature Photonics*, 10(6):420–427, 2016.

[281] X.-L. Qi and S.-C. Zhang. Topological insulators and superconductors. *Reviews of Modern Physics*, 83(4):1057, 2011.

[282] Y. Que, W. Xiao, H. Chen, D. Wang, S. Du, and H.-J. Gao. Stacking-dependent electronic property of trilayer graphene epitaxially grown on ru (0001). *Applied Physics Letters*, 107(26):263101, 2015.

[283] A. Rao, E. Richter, S. Bandow, B. Chase, P. Eklund, K. Williams, S. Fang, K. Subbaswamy, M. Menon, and A. Thess. Diameter-selective raman scattering from vibrational modes in carbon nanotubes. *Science*, 275(5297):187–191, 1997.

[284] A. M. Rao, P. Eklund, S. Bandow, A. Thess, and R. E. Smalley. Evidence for charge transfer in doped carbon nanotube bundles from raman scattering. *Nature*, 388(6639):257, 1997.

[285] R. Ravlić, M. Bode, A. Kubetzka, and R. Wiesendanger. Correlation of dislocation and domain structure of cr (001) investigated by spin-polarized scanning tunneling microscopy. *Physical Review B*, 67(17):174411, 2003.

[286] M. Ray, T. S. Basu, N. R. Bandyopadhyay, R. F. Klie, S. Ghosh, S. O. Raja, and A. K. Dasgupta. Highly lattice-mismatched semiconductor–metal hybrid nanostructures: gold nanoparticle encapsulated luminescent silicon quantum dots. *Nanoscale*, 6(4):2201–2210, 2014.

[287] B. Reed and M. Sarikaya. Electronic properties of carbon nanotubes by transmission electron energy-loss spectroscopy. *Physical Review B*, 64(19):195404, 2001.

[288] Z. Ren, Z. Huang, J. Xu, J. Wang, P. Bush, M. Siegal, and P. Provencio. Synthesis of large arrays of well-aligned carbon nanotubes on glass. *Science*, 282(5391):1105–1107, 1998.

[289] J.-W. Rhim and K. Moon. Edge states of zigzag bilayer graphite nanoribbons. *Journal of Physics: Condensed Matter*, 20(36):365202, 2008.

[290] K. A. Ritter and J. W. Lyding. The influence of edge structure on the electronic properties of graphene quantum dots and nanoribbons. *Nature Materials*, 8(3):235–242, 2009.

[291] D. Robertson, D. Brenner, and J. Mintmire. Energetics of nanoscale graphitic tubules. *Physical Review B*, 45(21):12592, 1992.

[292] J. T. Robinson, J. S. Burgess, C. E. Junkermeier, S. C. Badescu, T. L. Reinecke, F. K. Perkins, M. K. Zalalutdniov, J. W. Baldwin, J. C. Culbertson, and P. E. Sheehan. Properties of fluorinated graphene films. *Nano Letters*, 10(8):3001–3005, 2010.

[293] Z. Y. Rong and P. Kuiper. Electronic effects in scanning tunneling microscopy: Moiré pattern on a graphite surface. *Physical Review B*, 48(23):17427, 1993.

[294] P. Ruffieux, J. Cai, N. C. Plumb, L. Patthey, D. Prezzi, A. Ferretti, E. Molinari, X. Feng, K. Müllen, and C. A. Pignedoli. Electronic structure of atomically precise graphene nanoribbons. *ACS Nano*, 6(8):6930–6935, 2012.

[295] R. Rurali, V. Coluci, and D. Galvao. Prediction of giant electroactuation for papyruslike carbon nanoscroll structures: first-principles calculations. *Physical Review B*, 74(8):085414, 2006.

[296] S. Rusponi, N. Weiss, T. Cren, M. Epple, and H. Brune. High tunnel magnetoresistance in spin-polarized scanning tunneling microscopy of co nanoparticles on pt (111). *Applied Physics Letters*, 87(16):162514, 2005.

[297] C. Sabater, D. Gosálbez-Martínez, J. Fernández-Rossier, J. Rodrigo, C. Untiedt, and J. Palacios. Topologically protected quantum transport in locally exfoliated bismuth at room temperature. *Physical Review Letters*, 110(17):176802, 2013.

[298] R. Saito, M. Fujita, G. Dresselhaus, and M. S. Dresselhaus. Electronic structure of graphene tubules based on c 60. *Physical Review B*, 46(3):1804, 1992.

[299] R. Saito, M. Fujita, G. Dresselhaus, and u. M. Dresselhaus. Electronic structure of chiral graphene tubules. *Applied Physics Letters*, 60(18):2204–2206, 1992.

[300] M. Sano, A. Kamino, J. Okamura, and S. Shinkai. Ring closure of carbon nanotubes. *Science*, 293(5533):1299–1301, 2001.

[301] H. Santos, A. Ayuela, L. Chico, and E. Artacho. van der waals interaction in magnetic bilayer graphene nanoribbons. *Physical Review B*, 85(24):245430, 2012.

[302] B. Sarikavak-Lisesivdin, S. Lisesivdin, and E. Ozbay. Ab initio study of ru-terminated and ru-doped armchair graphene nanoribbons. *Molecular Physics*, 110(18):2295–2300, 2012.

[303] V. Saroka, M. Shuba, and M. Portnoi. Optical selection rules of zigzag graphene nanoribbons. *Physical Review B*, 95(15):155438, 2017.

[304] K.-i. Sasaki, K. Kato, Y. Tokura, K. Oguri, and T. Sogawa. Theory of optical transitions in graphene nanoribbons. *Physical Review B*, 84(8):085458, 2011.

[305] M. V. Savoskin, V. N. Mochalin, A. P. Yaroshenko, N. I. Lazareva, T. E. Konstantinova, I. V. Barsukov, and I. G. Prokofiev. Carbon nanoscrolls produced from acceptor-type graphite intercalation compounds. *Carbon*, 45(14):2797–2800, 2007.

[306] J. Schrier. Fluorinated and nanoporous graphene materials as sorbents for gas separations. *ACS Applied Materials & Interfaces*, 3(11):4451–4458, 2011.

[307] A. P. Seitsonen, A. M. Saitta, T. Wassmann, M. Lazzeri, and F. Mauri. Structure and stability of graphene nanoribbons in oxygen, carbon dioxide, water, and ammonia. *Physical Review B*, 82(11):115425, 2010.

[308] B. Senkovskiy, D. Haberer, D. Y. Usachov, A. Fedorov, N. Ehlen, M. Hell, L. Petaccia, G. Di Santo, R. Durr, F. Fischer, et al. Spectroscopic characterization of n= 9 armchair graphene nanoribbons. *Physica Status Solidi (RRL)-Rapid Research Letters*, 11(8), 2017.

[309] D. Shan, J. Zhang, H.-G. Xue, Y.-C. Zhang, S. Cosnier, and S.-N. Ding. Polycrystalline bismuth oxide films for development of amperometric biosensor for phenolic compounds. *Biosensors and Bioelectronics*, 24(12):3671–3676, 2009.

[310] D. B. Shinde, J. Debgupta, A. Kushwaha, M. Aslam, and V. K. Pillai. Electrochemical unzipping of multi-walled carbon nanotubes for facile synthesis of high-quality graphene nanoribbons. *Journal of the American Chemical Society*, 133(12):4168–4171, 2011.

[311] H. Shioyama and T. Akita. A new route to carbon nanotubes. *Carbon*, 41(1):179–181, 2003.

[312] M. Shlafman, T. Tabachnik, O. Shtempluk, A. Razin, V. Kochetkov, and Y. Yaish. Self aligned hysteresis free carbon nanotube field-effect transistors. *Applied Physics Letters*, 108(16):163104, 2016.

[313] F.-L. Shyu, C.-P. Chang, R.-B. Chen, C.-W. Chiu, and M. F. Lin. Magnetoelectronic and optical properties of carbon nanotubes. *Physical Review B*, 67(4):045405, 2003.

[314] F.-L. Shyu and M. F. Lin. Electronic and optical properties of narrow-gap carbon nanotubes. *Journal of the Physical Society of Japan*, 71(8):1820–1823, 2002.

[315] A. Sidorov, D. Mudd, G. Sumanasekera, P. Ouseph, C. Jayanthi, and S.-Y. Wu. Electrostatic deposition of graphene in a gaseous environment: a deterministic route for synthesizing rolled graphenes? *Nanotechnology*, 20(5):055611, 2009.

[316] D. A. Siegel, W. Regan, A. V. Fedorov, A. Zettl, and A. Lanzara. Charge-carrier screening in single-layer graphene. *Physical Review Letters*, 110(14):146802, 2013.

[317] A. J. Simbeck, D. Gu, N. Kharche, P. V. Satyam, P. Avouris, and S. K. Nayak. Electronic structure of oxygen-functionalized armchair graphene nanoribbons. *Physical Review B*, 88(3):035413, 2013.

[318] K. A. Simonov, N. A. Vinogradov, A. S. Vinogradov, A. V. Generalov, G. I. Svirskiy, A. A. Cafolla, N. Mårtensson, and A. B. Preobrajenski. Effect of electron injection in copper-contacted graphene nanoribbons. *Nano Research*, 9(9):2735–2746, 2016.

[319] K. Singh and B. Raj. Temperature-dependent modeling and performance evaluation of multi-walled cnt and single-walled cnt as global interconnects. *Journal of Electronic Materials*, 44(12):4825, 2015.

[320] B. W. Smith, M. Monthioux, and D. E. Luzzi. Encapsulated c60 in carbon nanotubes. *Nature*, 396(6709):323, 1998.

[321] H. Söde, L. Talirz, O. Gröning, C. A. Pignedoli, R. Berger, X. Feng, K. Müllen, R. Fasel, and P. Ruffieux. Electronic band dispersion of graphene nanoribbons via fourier-transformed scanning tunneling spectroscopy. *Physical Review B*, 91(4):045429, 2015.

[322] E. Sohmen, J. Fink, and W. Krätschmer. Electronic structure studies of undoped and n-type doped fullerene c60. *EPL (Europhysics Letters)*, 17(1):51, 1992.

[323] Y.-W. Son, M. L. Cohen, and S. G. Louie. Energy gaps in graphene nanoribbons. *Physical Review Letters*, 97(21):216803, 2006.

[324] Y.-W. Son, M. L. Cohen, and S. G. Louie. Half-metallic graphene nanoribbons. *Nature*, 444(7117):347–349, 2006.

[325] M. Sprinkle, M. Ruan, Y. Hu, J. Hankinson, M. Rubio-Roy, B. Zhang, X. Wu, C. Berger, and W. A. De Heer. Scalable templated growth of graphene nanoribbons on sic. *Nature Nanotechnology*, 5(10):727–731, 2010.

[326] E. Stolyarova, D. Stolyarov, K. Bolotin, S. Ryu, L. Liu, K. Rim, M. Klima, M. Hybertsen, I. Pogorelsky, and I. Pavlishin. Observation of graphene bubbles and effective mass transport under graphene films. *Nano Letters*, 9(1):332–337, 2008.

[327] K. Sugawara, T. Sato, S. Souma, T. Takahashi, and H. Suematsu. Fermi surface and edge-localized states in graphite studied by high-resolution angle-resolved photoemission spectroscopy. *Physical Review B*, 73(4):045124, 2006.

[328] C. Sun, Y. Feng, Y. Li, C. Qin, Q. Zhang, and W. Feng. Solvothermally exfoliated fluorographene for high-performance lithium primary batteries. *Nanoscale*, 6(5):2634–2641, 2014.

[329] P. Sutter, M. Hybertsen, J. Sadowski, and E. Sutter. Electronic structure of few-layer epitaxial graphene on ru (0001). *Nano Letters*, 9(7):2654–2660, 2009.

[330] L. Talirz, H. Söde, J. Cai, P. Ruffieux, S. Blankenburg, R. Jafaar, R. Berger, X. Feng, K. Müllen, and D. Passerone. Termini of bottom-up fabricated graphene nanoribbons. *Journal of the American Chemical Society*, 135(6):2060–2063, 2013.

[331] L. Talirz, H. Sode, T. Dumslaff, S. Wang, P. Shinde, C. A. Pignedoli, L. Liang, and V. Meunier. On-surface synthesis and characterization of 9-atom wide armchair graphene nanoribbons. *ACS Nano*, 11(2):1380–1388, 2017.

[332] A. V. Talyzin, S. Luzan, I. V. Anoshkin, A. G. Nasibulin, H. Jiang, E. I. Kauppinen, V. M. Mikoushkin, V. V. Shnitov, D. E. Marchenko, and D. Noreus. Hydrogenation, purification, and unzipping of carbon nanotubes by reaction with molecular hydrogen: road to graphane nanoribbons. *ACS Nano*, 5(6):5132–5140, 2011.

[333] Y.-Z. Tan, B. Yang, K. Parvez, A. Narita, S. Osella, D. Beljonne, X. Feng, and K. Müllen. Atomically precise edge chlorination of nanographenes and its application in graphene nanoribbons. *Nature Communications*, 4:2646.

[334] S. J. Tans, A. R. Verschueren, and C. Dekker. Room-temperature transistor based on a single carbon nanotube. *Nature*, 393(6680):49–52, 1998.

[335] C. Tao, L. Jiao, O. V. Yazyev, Y.-C. Chen, J. Feng, X. Zhang, R. B. Capaz, J. M. Tour, A. Zettl, and S. G. Louie. Spatially resolving edge states of chiral graphene nanoribbons. *Nature Physics*, 7(8):616–620, 2011.

[336] L. Tapasztó, G. Dobrik, P. Lambin, and L. P. Biró. Tailoring the atomic structure of graphene nanoribbons by scanning tunnelling microscope lithography. *Nature Nanotechnology*, 3(7):397–401, 2008.

[337] A. Thess, R. Lee, P. Nikolaev, H. Dai, P. Petit, J. Robert, C. Xu, Y. H. Lee, S. G. Kim, A. G. Rinzler, et al. Crystalline ropes of metallic carbon nanotubes. *Science*, 273(5274):483–487, 1996.

[338] D. Tomanek, W. Zhong, and E. Krastev. Stability of multishell fullerenes. *Physical Review B*, 48(20):15461, 1993.

[339] N. T. T. Tran, S.-Y. Lin, O. E. Glukhova, and M. F. Lin. Configuration-induced rich electronic properties of bilayer graphene. *The Journal of Physical Chemistry C*, 119(19):10623–10630, 2015.

[340] N. T. T. Tran, S. Y. Lin, C. Y. Lin, and M. F. Lin. *Geometric and Electronic Properties of Graphene-Related Systems: Chemical Bondings.* CRC Press, Boca Raton, FL, 2017.

[341] F. Traversi, C. Raillon, S. Benameur, K. Liu, S. Khlybov, M. Tosun, D. Krasnozhon, A. Kis, and A. Radenovic. Detecting the translocation of dna through a nanopore using graphene nanoribbons. *Nature Nanotechnology*, 8(12):939–945, 2013.

[342] F. Tuinstra and J. L. Koenig. Raman spectrum of graphite. *The Journal of Chemical Physics*, 53(3):1126–1130, 1970.

[343] V. Urbanová, F. Karlický, A. Matěj, F. Šembera, Z. Janoušek, J. A. Perman, V. Ranc, K. Čépe, J. Michl, and M. Otyepka. Fluorinated graphenes as advanced biosensors–effect of fluorine coverage on electron transfer properties and adsorption of biomolecules. *Nanoscale*, 8(24):12134–12142, 2016.

[344] D. Usachov, A. Fedorov, M. M. Otrokov, A. Chikina, O. Vilkov, A. Petukhov, A. G. Rybkin, and V. K. Adamchuk. Observation of single-spin dirac fermions at the graphene/ferromagnet interface. *Nano Letters*, 15(4):2396–2401, 2015.

[345] L. M. Viculis, J. J. Mack, and R. B. Kaner. A chemical route to carbon nanoscrolls. *Science*, 299(5611):1361–1361, 2003.

[346] B. Vinayan, R. Nagar, V. Raman, N. Rajalakshmi, K. Dhathathreyan, and S. Ramaprabhu. Synthesis of graphene-multiwalled carbon nanotubes hybrid nanostructure by strengthened electrostatic interaction and its lithium ion battery application. *Journal of Materials Chemistry*, 22(19):9949–9956, 2012.

[347] C. Virojanadara, S. Watcharinyanon, A. Zakharov, and L. I. Johansson. Epitaxial graphene on 6 h-sic and li intercalation. *Physical Review B*, 82(20):205402, 2010.

[348] P. Wagner, C. P. Ewels, J.-J. Adjizian, L. Magaud, P. Pochet, S. Roche, A. Lopez-Bezanilla, V. V. Ivanovskaya, A. Yaya, and M. Rayson. Band gap engineering via edge-functionalization of graphene nanoribbons. *The Journal of Physical Chemistry C*, 117(50):26790–26796, 2013.

[349] K. Wakabayashi, M. Fujita, H. Ajiki, and M. Sigrist. Electronic and magnetic properties of nanographite ribbons. *Physical Review B*, 59(12):8271, 1999.

[350] K. Wakabayashi, Y. Takane, M. Yamamoto, and M. Sigrist. Electronic transport properties of graphene nanoribbons. *New Journal of Physics*, 11(9):095016, 2009.

[351] P. R. Wallace. The band theory of graphite. *Physical Review*, 71(9):622, 1947.

[352] A. L. Walter, K.-J. Jeon, A. Bostwick, F. Speck, M. Ostler, T. Seyller, L. Moreschini, Y. S. Kim, Y. J. Chang, and K. Horn. Highly p-doped epitaxial graphene obtained by fluorine intercalation. *Applied Physics Letters*, 98(18):184102, 2011.

[353] A. K. Wanekaya. Applications of nanoscale carbon-based materials in heavy metal sensing and detection. *Analyst*, 136(21):4383–4391, 2011.

[354] S. Wang, Q. Zeng, L. Yang, Z. Zhang, Z. Wang, T. Pei, L. Ding, X. Liang, M. Gao, and Y. Li. High-performance carbon nanotube light-emitting diodes with asymmetric contacts. *Nano Letters*, 11(1):23–29, 2010.

[355] X. Wang, Y. Ouyang, X. Li, H. Wang, J. Guo, and H. Dai. Room-temperature all-semiconducting sub-10-nm graphene nanoribbon field-effect transistors. *Physical Review Letters*, 100(20):206803, 2008.

[356] Y. Wang, C. Cao, and H.-P. Cheng. Metal-terminated graphene nanoribbons. *Physical Review B*, 82(20):205429, 2010.

[357] Y. Wang, Z. Ni, L. Liu, Y. Liu, C. Cong, T. Yu, X. Wang, D. Shen, and Z. Shen. Stacking-dependent optical conductivity of bilayer graphene. *ACS Nano*, 4(7):4074–4080, 2010.

[358] Z. Wang, M.-Y. Yao, W. Ming, L. Miao, F. Zhu, C. Liu, C. Gao, D. Qian, J.-F. Jia, and F. Liu. Creation of helical dirac fermions by interfacing two gapped systems of ordinary fermions. *Nature Communications*, 4:1384, 2013.

[359] J. H. Warner, M. H. Rümmeli, T. Gemming, B. Büchner, and G. A. D. Briggs. Direct imaging of rotational stacking faults in few layer graphene. *Nano Letters*, 9(1):102–106, 2008.

[360] K. Watanabe, T. Kondow, M. Soma, T. Onishi, and K. Tamaru. Molecular-sieve type sorption on alkali graphite intercalation compounds. *Proceedings of the Royal Society of London A: Mathematical, Physical and Engineering Sciences*, 333(1592):51–67, 1973.

[361] D. Wei, Y. Liu, H. Zhang, L. Huang, B. Wu, J. Chen, and G. Yu. Scalable synthesis of few-layer graphene ribbons with controlled morphologies by a template method and their applications in nanoelectromechanical switches. *Journal of the American Chemical Society*, 131(31):11147–11154, 2009.

[362] R. Wiesendanger. Spin mapping at the nanoscale and atomic scale. *Reviews of Modern Physics*, 81(4):1495, 2009.

[363] R. Wiesendanger, H.-J. Güntherodt, G. Güntherodt, R. Gambino, and R. Ruf. Observation of vacuum tunneling of spin-polarized electrons with the scanning tunneling microscope. *Physical Review Letters*, 65(2):247, 1990.

[364] J. W. Wilder, L. C. Venema, A. G. Rinzler, R. E. Smalley, and C. Dekker. Electronic structure of atomically resolved carbon nanotubes. *Nature*, 391(6662):59–62, 1998.

[365] F. Withers, S. Russo, M. Dubois, and M. F. Craciun. Tuning the electronic transport properties of graphene through functionalisation with fluorine. *Nanoscale Research Letters*, 6(1):526, 2011.

[366] S. Wolf, D. Awschalom, R. Buhrman, J. Daughton, S. Von Molnar, M. Roukes, A. Y. Chtchelkanova, and D. Treger. Spintronics: a spin-based electronics vision for the future. *Science*, 294(5546):1488–1495, 2001.

[367] B.-R. Wu. Field modulation of the electronic structure of trilayer graphene. *Applied Physics Letters*, 98(26):263107, 2011.

[368] J.-Y. Wu, S.-C. Chen, O. Roslyak, G. Gumbs, and M. F. Lin. Plasma excitations in graphene: Their spectral intensity and temperature dependence in magnetic field. *ACS Nano*, 5(2):1026–1032, 2011.

[369] L. Xie, H. Wang, C. Jin, X. Wang, L. Jiao, K. Suenaga, and H. Dai. Graphene nanoribbons from unzipped carbon nanotubes: atomic structures, raman spectroscopy, and electrical properties. *Journal of the American Chemical Society*, 133(27):10394–10397, 2011.

[370] X. Xie, L. Ju, X. Feng, Y. Sun, R. Zhou, K. Liu, S. Fan, Q. Li, and K. Jiang. Controlled fabrication of high-quality carbon nanoscrolls from monolayer graphene. *Nano Letters*, 9(7):2565–2570, 2009.

[371] Y. E. Xie, Y. P. Chen, and J. Zhong. Electron transport of folded graphene nanoribbons. *Journal of Applied Physics*, 106(10):103714, 2009.

[372] Y. Xing, Q.-f. Sun, and J. Wang. Nernst and seebeck effects in a graphene nanoribbon. *Physical Review B*, 80(23):235411, 2009.

[373] P. Xu, Y. Yang, D. Qi, S. Barber, J. Schoelz, M. Ackerman, L. Bellaiche, and P. Thibado. Electronic transition from graphite to graphene via controlled movement of the top layer with scanning tunneling microscopy. *Physical Review B*, 86(8):085428, 2012.

[374] J. Yan, C. Li, D. Zhan, L. Liu, D. Shen, J.-L. Kuo, S. Chen, and Z. Shen. Graphene homojunction: closed-edge bilayer graphene by pseudospin interaction. *Nanoscale*, 8(17):9102–9106, 2016.

[375] F. Yang, L. Miao, Z. Wang, M.-Y. Yao, F. Zhu, Y. Song, M.-X. Wang, J.-P. Xu, A. V. Fedorov, and Z. Sun. Spatial and energy distribution of topological edge states in single bi (111) bilayer. *Physical Review Letters*, 109(1):016801, 2012.

[376] L. Yang, M. L. Cohen, and S. G. Louie. Excitonic effects in the optical spectra of graphene nanoribbons. *Nano Letters*, 7(10):3112–3115, 2007.

[377] X. Yang, X. Dou, A. Rouhanipour, L. Zhi, H. J. Räder, and K. Müllen. Two-dimensional graphene nanoribbons. *Journal of the American Chemical Society*, 130(13):4216–4217, 2008.

[378] M. Yankowitz, F. Wang, C. N. Lau, and B. J. LeRoy. Local spectroscopy of the electrically tunable band gap in trilayer graphene. *Physical Review B*, 87(16):165102, 2013.

[379] Y. Yayon, V. Brar, L. Senapati, S. Erwin, and M. Crommie. Observing spin polarization of individual magnetic adatoms. *Physical Review Letters*, 99(6):067202, 2007.

[380] O. V. Yazyev and M. Katsnelson. Magnetic correlations at graphene edges: basis for novel spintronics devices. *Physical Review Letters*, 100(4):047209, 2008.

[381] M.-Y. Yen, M.-C. Hsiao, S.-H. Liao, P.-I. Liu, H.-M. Tsai, C.-C. M. Ma, N.-W. Pu, and M.-D. Ger. Preparation of graphene/multi-walled carbon nanotube hybrid and its use as photoanodes of dye-sensitized solar cells. *Carbon*, 49(11):3597–3606, 2011.

[382] W.-J. Yin, Y.-E. Xie, L.-M. Liu, Y.-P. Chen, R.-Z. Wang, X.-L. Wei, J.-X. Zhong, and L. Lau. Atomic structure and electronic properties of folded graphene nanoribbons: A first-principles study. *Journal of Applied Physics*, 113(17):173506, 2013.

[383] D. Yu and L. Dai. Self-assembled graphene/carbon nanotube hybrid films for supercapacitors. *The Journal of Physical Chemistry Letters*, 1(2):467–470, 2009.

[384] S. Yuan, R. Roldán, and M. I. Katsnelson. Excitation spectrum and high-energy plasmons in single-layer and multilayer graphene. *Physical Review B*, 84(3):035439, 2011.

[385] R. Yuksel, Z. Sarioba, A. Cirpan, P. Hiralal, and H. E. Unalan. Transparent and flexible supercapacitors with single walled carbon nanotube thin film electrodes. *ACS Applied Materials & Interfaces*, 6(17):15434–15439, 2014.

[386] S. Zaric, G. N. Ostojic, J. Kono, J. Shaver, V. C. Moore, M. S. Strano, R. H. Hauge, R. E. Smalley, and X. Wei. Optical signatures of the aharonov-bohm phase in single-walled carbon nanotubes. *Science*, 304(5674):1129–1131, 2004.

[387] J. Zhang, J. Xiao, X. Meng, C. Monroe, Y. Huang, and J.-M. Zuo. Free folding of suspended graphene sheets by random mechanical stimulation. *Physical Review Letters*, 104(16):166805, 2010.

[388] T.-F. Zhang, Z.-P. Li, J.-Z. Wang, W.-Y. Kong, G.-A. Wu, Y.-Z. Zheng, Y.-W. Zhao, E.-X. Yao, N.-X. Zhuang, and L.-B. Luo. Broadband photodetector based on carbon nanotube thin film/single layer graphene schottky junction. *Scientific Reports*, 6:38569, 2016.

[389] X. Zhang and G. Lu. The spin-orbit coupling induced spin flip and its role in the enhancement of the photocatalytic hydrogen evolution over iodinated graphene oxide. *Carbon*, 108:215–224, 2016.

[390] X. Zhang, O. V. Yazyev, J. Feng, L. Xie, C. Tao, Y.-C. Chen, L. Jiao, Z. Pedramrazi, A. Zettl, and S. G. Louie. Experimentally engineering the edge termination of graphene nanoribbons. *ACS Nano*, 7(1):198–202, 2012.

[391] Y. Zhang, Y.-W. Tan, H. L. Stormer, and P. Kim. Experimental observation of the quantum hall effect and berry's phase in graphene. *Nature*, 438(7065):201, 2005.

[392] Z. Zhang and W. Guo. Tunable ferromagnetic spin ordering in boron nitride nanotubes with topological fluorine adsorption. *Journal of the American Chemical Society*, 131(19):6874–6879, 2009.

[393] F.-G. Zhao, G. Zhao, X.-H. Liu, C.-W. Ge, J.-T. Wang, B.-L. Li, Q.-G. Wang, W.-S. Li, and Q.-Y. Chen. Fluorinated graphene: facile solution preparation and tailorable properties by fluorine-content tuning. *J. Mater. Chem. A*, 2(23):8782–8789, 2014.

[394] X. Zhao, Y. Ando, Y. Liu, M. Jinno, and T. Suzuki. Carbon nanowire made of a long linear carbon chain inserted inside a multiwalled carbon nanotube. *Physical Review Letters*, 90(18):187401, 2003.

[395] H. Zheng, Z. Wang, T. Luo, Q. Shi, and J. Chen. Analytical study of electronic structure in armchair graphene nanoribbons. *Phys. Rev. B*, 75(16):165414, 2007.

[396] X. Zheng, L. Song, R. Wang, H. Hao, L. Guo, and Z. Zeng. Electronic structures and transverse electrical field effects in folded zigzag-edged graphene nanoribbons. *Applied Physics Letters*, 97(15):153129, 2010.

[397] H. Zhong, K. Xu, Z. Liu, G. Xu, L. Shi, Y. Fan, J. Wang, G. Ren, and H. Yang. Charge transport mechanisms of graphene/semiconductor schottky barriers: A theoretical and experimental study. *J. Appl. Phys.*, 115(1):013701, 2014.

[398] S. Zhou, D. Siegel, A. Fedorov, and A. Lanzara. Metal to insulator transition in epitaxial graphene induced by molecular doping. *Physical Review Letters*, 101(8):086402, 2008.

[399] F.-f. Zhu, W.-j. Chen, Y. Xu, C.-l. Gao, D.-d. Guan, C.-h. Liu, D. Qian, S.-C. Zhang, and J.-f. Jia. Epitaxial growth of two-dimensional stanene. *Nature Materials*, 14(10):1020–1025, 2015.

# Index

π bondings
  alkali-adsorbed nanoribbons, 154
  aluminum-adsorbed
    nanoribbons, 198
  fluorine-adsorbed nanoribbons,
    171
  halogen-adsorbed nanoribbons,
    183
σ bondings
  alkali-adsorbed nanoribbons, 154
  aluminum-adsorbed
    nanoribbons, 198
  fluorine-adsorbed nanoribbons,
    171
  halogen-adsorbed nanoribbons,
    183

adatom, 124, 163
adsorption energies
  edge-decorated nanoribbons, 130
AFM (atomic force microscopy), 16
alkali-adsorbed nanoribbons, 147
  π bondings, 154
  σ bondings, 154
  ARPES, 159
  binding energy, 148
  charge differences
    armchair, 154
  charge distributions
    armchair, 154
  DOS
    armchair, 154
    zigzag, 159
  energy bands
    armchair, 148–154
    zigzag, 157
  Fermi level, 148

geometric structures
  armchair, 148
  introduction to, 5
  lithium, 148
  n-type doping, 151
  overview of, 238
  sodium, 148
  spin densities, 157
aluminum-adsorbed nanoribbon
  charge differences, 198
  charge distributions, 198
  DOS, 198
  energy bands
    armchair, 192
    zigzag, 195
  geometric structures, 192
  n-type doping, 195
  spin densities, 195
aluminum-adsorbed nanoribbons
  π bondings, 198
  σ bondings, 198
  multi-orbtial hybridizations, 209
anti-ferromagnetism
  bismuth-adsorbed nanoribbon,
    217
  curved decorated nanoribbons,
    143
  fluorine-adsorbed nanoribbons,
    169
  halogen-adsorbed nanoribbons,
    180
  monolayer nanoribbons, 25
    zigzag, 39
  titanium-adsorbed nanoribbon,
    206
arc angle

curved decorated nanoribbons, 136
curved nanoribbons, 50
ARPES, 21–23
    alkali-adsorbed nanoribbons, 159
    halogen-adsorbed nanoribbons, 185
    monolayer nanoribbons, 46

Bader analysis, 149
    edge-decorated nanoribbons, 124
bilayer nanoribbons
    cohesive energy, 107
    DOS, 112–114, 117
    energy bands, 110–112, 117
    geometric structures, 104
    introduction to, 4
    magnetism, 104, 107
    overview of, 236–237
    semiconductor-metal transition, 112
    STM, 103
    van Hove singularity, 117
binding energy
    alkali-adsorbed nanoribbons, 148
bismuth-adsorbed nanoribbon
    anti-ferromagnetism, 217
    charge distributions, 219
    DOS, 222
    energy bands
        armchair, 212, 215
        zigzag, 217
    ferromagnetism, 217
    geometric structures, 212
    spin densities
        armchair, 217
        zigzag, 219
bismuth-adsorbed nanoribbons
    charge differences, 219
    multi-orbtial hybridizations, 219

carbon nanoscrolls
    charge distributions, 94
    critical widths, 91
    DOS, 94

energy bands, 93–94
    armchair, 93
    zigzag, 94
energy gaps, 94
formation energy, 88
geometric structures, 88–93
introduction to, 3
overview of, 236
syntheses, 87
carbon nanotubes
    applications, 69
    armchair nanotube, 61
    composite system, 65
    DOS, 65
    energy gaps, 63–65
    geometric structures, 59
    magnetic flux, 63
    spatial charge distribution, 63
    tight-binding model, 59–61
    unzip, 123
    zigzag nanotube, 61–63
charge differences
    alkali-adsorbed nanoribbons
        armchair, 154
    aluminum-adsorbed nanoribbon, 198
    bismuth-adsorbed nanoribbons, 219
    fluorine-adsorbed nanoribbons, 171
    halogen-adsorbed nanoribbons, 183
    monolayer nanoribbons
        armchair, 33
        zigzag, 39
    titanium-adsorbed nanoribbon, 209
charge distributions
    alkali-adsorbed nanoribbons
        armchair, 154
    aluminum-adsorbed nanoribbon, 198
    bismuth-adsorbed nanoribbon, 219
    carbon nanoscrolls, 94

edge-decorated nanoribbons
  nitrogen, 125
fluorine-adsorbed nanoribbons,
    171
halogen-adsorbed nanoribbons,
    183
monolayer nanoribbons
  armchair, 33
  zigzag, 39
titanium-adsorbed nanoribbon,
    209
chlorination, 6
cohesive energy
  bilayer nanoribbons, 107
curved decorated nanoribbons
  anti-ferromagnetism, 143
  arc angle, 136
  beryllium, 136
  boron, 136
  carbon, 136
  critical angle, 136
  DOS, 143–145
  energy bands, 142–143
  geometric structures, 134–142
  hydrogen, 136
  introduction to, 5
  magnesium, 138
  scandium, 142
  titanium, 140
  total energy, 136
  transition metal, 140
  type I, 136
  type II, 138
  type III, 140
  zinc, 142
curved graphene nanoribbons
  critical angle, 52
  DOS, 56–57
  energy bands, 52–56
  semiconductor-metal transition,
      55
  STM, 57
  total energy, 52
curved nanoribbons
  arc angle, 50

geometric structures, 50
introduction to, 2
overview of, 234
synthesis, 49
total energy, 50

de Broglie wavelength, 15
DOS
  alkali-adsorbed nanoribbons
    armchair, 154
    zigzag, 159
  aluminum-adsorbed nanoribbon,
      198
  bilayer nanoribbons, 112–114
  bismuth-adsorbed nanoribbon,
      222
  carbon nanoscrolls, 94
  carbon nanotube, 65
  curved decorated nanoribbons,
      143–145
  curved graphene nanoribbons,
      56–57
  edge-decorated nanoribbons,
      132–134
  Fe/Co/Ni-adsorbed nanoribbon,
      227
  fluorine-adsorbed nanoribbons,
      173
  folded nanoribbons, 84
  halogen-adsorbed nanoribbons,
      183
  monolayer nanoribbons
    armchair, 33
    zigzag, 43
  titanium-adsorbed nanoribbon,
      209

edge passivation
  boron, 123
  fluorine, 123
  magnesium, 123
  oxygen, 123
  potassium, 123
  ruthenium, 123
  tellurium, 123

transition metal, 123
edge-decorated nanoribbons
   adsorption energies, 130
   Bader analysis, 124
   charge distributions
      nitrogen, 125
   DOS, 132–134
   energy bands, 130–132
   energy gaps, 132
   geometric structures, 124–130
      aluminum, 128
      beryllium, 125
      boron, 125
      calcium, 128
      chlorine, 128
      fluorine, 128
      lithium, 128
      magnesium, 128
      nitrogen, 125
      oxygen, 128
      phosphorus, 128
      potassium, 128
      silicon, 128
      sodium, 128
      sulfur, 128
   introduction to, 4
   overview of, 237–238
   semiconductor-metal transition,
      132
   STM, 123
energy bands, 130
   alkali-adsorbed nanoribbons
      armchair, 148–154
      zigzag, 157
   aluminum-adsorbed nanoribbon
      armchair, 192
      zigzag, 195
   armchair nanotube, 61
   bilayer nanoribbons, 110–112
   bismuth-adsorbed nanoribbon
      armchair, 212, 215
      zigzag, 217
   carbon nanoscrolls, 93–94
      armchair, 93
      zigzag, 94

   curved decorated nanoribbons,
      142–143
   curved graphene nanoribbons,
      52–56
   edge-decorated nanoribbons,
      130–132
   Fe/Co/Ni-adsorbed nanoribbon
      armchair, 225
      zigzag, 227
   fluorine-adsorbed nanoribbons,
      164–171
   folded nanoribbons, 77, 80–83
   halogen-adsorbed nanoribbons,
      176–177
   titanium-adsorbed nanoribbon
      armchair, 203
      zigzag, 203
   zigzag nanotube, 61–63
energy gaps
   carbon nanoscrolls, 94
   edge-decorated nanoribbons, 132
   folded nanoribbons
      armchair, 77
      zigzag, 83

Fe/Co/Ni-adsorbed nanoribbon
   DOS, 227
      energy bands
         armchair, 225
         zigzag, 227
      geometric structures, 225
Fermi level
   alkali-adsorbed nanoribbons, 148
ferromagnetism
   bismuth-adsorbed nanoribbon,
      217
   fluorine-adsorbed nanoribbons,
      169
   halogen-adsorbed nanoribbons,
      180
   titanium-adsorbed nanoribbon,
      206
first-principles, 11
fluorination, 6
fluorine-adsorbed nanoribbons

$\pi$ bondings, 171
$\sigma$ bondings, 171
anti-ferromagnetism, 169
charge differences, 171
charge distributions, 171
DOS, 173
energy bands, 164–171
ferromagnetism, 169
geometric structures, 164
p-type doping, 164, 167, 169
spin densities, 171
STS, 176
folded nanoribbons
  DOS, 84
  energy bands, 77, 80–83
  energy gaps
    armchair, 77
    zigzag, 83
  experiments, 71
  folding energy, 74–77
  geometric structures, 72–74
  introduction to, 3
  magnetism, 80
  overview of, 235–236
  SP-STS, 84
  STS, 84
  van Hove singularity, 84
  width-dependence, 74

geometric structures
  alkali-adsorbed nanoribbons
    armchair, 148
  aluminum-adsorbed nanoribbon, 192
  bilayer nanoribbons, 104
  bismuth-adsorbed nanoribbon, 212
  carbon nanoscrolls, 88–93
  carbon nanotubes, 59
  curved decorated nanoribbons, 134–142
  curved nanoribbons, 50
  edge-decorated nanoribbons, 124–130
    aluminum, 128

beryllium, 125
boron, 125
calcium, 128
chlorine, 128
fluorine, 128
lithium, 128
magnesium, 128
nitrogen, 125
oxygen, 128
phosphorus, 128
potassium, 128
silicon, 128
sodium, 128
sulfur, 128
Fe/Co/Ni-adsorbed nanoribbon, 225
folded nanoribbons, 72–74
titanium-adsorbed nanoribbon, 201

halogen-adsorbed nanoribbons
  $\pi$ bondings, 183
  $\sigma$ bondings, 183
  anti-ferromagnetism, 180
  ARPES, 185
  charge differences, 183
  charge distributions, 183
  DOS, 183
  energy bands, 176–177
  ferromagnetism, 180
  geometric structures, 176
  introduction to, 6
  overview of, 238–239
  p-type doping, 177
  semiconductor-metal transition, 185
  STS, 185

Kohn-Sham equations, 11

magnetism
  bilayer nanoribbons, 104, 107
  folded nanoribbons, 80
metal-adsorbed nanoribbons, 189
  ARPES, 189
  introduction to, 6

n-type doping, 189
overview of, 239–240
monolayer nanoribbons
    anti-ferromagnetism, 25, 39
    ARPES, 26, 46
    charge differences
        armchair, 33
        zigzag, 39
    charge distributions
        armchair, 33
        zigzag, 39
    DOS
        armchair, 33
        zigzag, 43
    energy bands
        armchair, 27
        zigzag, 37
    energy gaps
        armchair, 29
        zigzag, 39
    geometric structures
        armchair, 27
        zigzag, 35
    Hubbard model, 33
    introduction to, 2
    Landau subbands, 47
    overview of, 233
    STM, 25, 26
    STS, 26, 43
    tight-binding model, 25
        armchair, 29
multi-orbtial hybridizations
    aluminum-adsorbed
        nanoribbons, 209
    bismuth-adsorbed nanoribbons,
        219

n-type doping
    aluminum-adsorbed nanoribbon,
        195
    metal-adsorbed nanoribbons,
        189

p-type doping
    fluorine-adsorbed nanoribbons,
        164, 167, 169

halogen-adsorbed nanoribbons,
    177

quantum confinement, 12

semiconductor-metal transition, 3,
    83, 103
    bilayer nanoribbons, 112
    edge-decorated nanoribbons, 132
    halogen-adsorbed nanoribbons,
        185
SP-STM, 20–21
SP-STS, 21, 99
spin densities
    alkali-adsorbed nanoribbons, 157
    aluminum-adsorbed nanoribbon,
        195
    bismuth-adsorbed nanoribbon
        armchair, 217
        zigzag, 219
    fluorine-adsorbed nanoribbons,
        171
    titanium-adsorbed nanoribbon
        armchair, 206
        zigzag, 206
STM, 15–16
    bilayer nanoribbons, 103
    edge-decorated nanoribbons, 123
STS, 16–20
    halogen-adsorbed nanoribbons,
        185
    monolayer nanoribbons, 43

TEM, 14–15
tight-binding model
    monolayer nanoribbons, 25
titanium-adsorbed nanoribbon
    anti-ferromagnetism, 206
    charge differences, 209
    charge distributions, 209
    DOS, 209
    energy bands
        armchair, 203
        zigzag, 203
    ferromagnetism, 206
    geometric structures, 201

spin densities
   armchair, 206
   zigzag, 206
van Hove singularity, 209
transition metal
   curved decorated nanoribbons,
      140

unzip
   carbon nanotubes, 123

van Der Walls, 3, 12, 65, 71, 74, 104,
      191
van Hove singularity
   folded nanoribbons, 84
   titanium-adsorbed nanoribbon,
      209